Hume and the Demands of Philosophy

Hume and the Demands of Philosophy

Science, Skepticism, and Moderation

Nathan I. Sasser

LEXINGTON BOOKS
Lanham • Boulder • New York • London

Published by Lexington Books
An imprint of The Rowman & Littlefield Publishing Group, Inc.
4501 Forbes Boulevard, Suite 200, Lanham, Maryland 20706
www.rowman.com

86-90 Paul Street, London EC2A 4NE

British Library Cataloguing in Publication Information Available

Library of Congress Cataloging-in-Publication Data

Names: Sasser, Nathan I., author.
Title: Hume and the demands of philosophy : science, skepticism, and moderation / Nathan I. Sasser.
Description: Lanham : Lexington Books, 2022. | Includes bibliographical references and index. | Summary: "This book argues that Hume is a radical epistemic skeptic who gives only practical reasons for retaining belief in sensory beliefs and the deliverances of reason. He advises us to take a moderate approach to the demands of philosophy, since they sometimes diverge from the demands of life"— Provided by publisher.
Identifiers: LCCN 2021053143 (print) | LCCN 2021053144 (ebook) | ISBN 9781793623218 (cloth) | ISBN 9781793623225 (epub)
Subjects: LCSH: Hume, David, 1711-1776.
Classification: LCC B1498 .S27 2022 (print) | LCC B1498 (ebook) | DDC 192—dc23/eng/20211108
LC record available at https://lccn.loc.gov/2021053143
LC ebook record available at https://lccn.loc.gov/2021053144

Contents

Acknowledgments

It is a great pleasure to thank the many people who have contributed in various ways to the emergence of this book. My mother and father introduced me to the life of the mind and remain some of my favorite and most regular interlocutors. John Carroll, my undergraduate philosophy professor and advisor at North Carolina State University, provided a formative example of rigorous analytic philosophy, as well as personal encouragement. Jared Oliphint has been a philosophical fellow traveler for almost twenty years; I'm glad we're both still on the road. Whatever I know about writing in the history of philosophy is due to Matt Kisner. Hours of conversation with him helped me clarify my early thoughts about Hume.

I've been honored to have some of the Hume scholars I most respect critique earlier drafts of this material. Don Garrett has been exceptionally generous in reading and responding to earlier versions of much of this book. I have learned so much from him (though not as much as I should have). Hsueh Qu also has been kind enough to provide critical comments on several chapters. Through the Hume Society's mentoring program, I have received valuable feedback and professional advice from Donald Ainslie, James Harris, and David Owen. An anonymous reviewer for Lexington Books provided world-class critical comments on the entire manuscript of this book. I bear full responsibility for the remaining defects of this book, but they would have been far more egregious without the input of these scholars.

I owe a large debt of gratitude to both the Institute for the Study of Scottish Philosophy (formerly the Center for the Study of Scottish Philosophy) and the Hume Society, whose constituencies happily overlap. Their respective annual conferences are delightful and invigorating. Over the years I have profited from conversations with Miren Boehm, James Foster, Giovanni Gellera, Gordon Graham, Mike Jacovides, Emily Kelahan, Esther Kroeker,

Doug McDermid, Ben Nelson, Taro Okamura, Phil Reed, Todd Ryan, Corliss Swain, and many more scholars than I can list here.

The local South Carolina philosophical community has sustained me in various ways. The South Carolina Society for Philosophy keeps me connected to philosophers from across the state and their wide-ranging research. I am grateful for the friendship and professional support of several philosophers at Furman University (including David Gandolfo, Aaron Simmons, Mark Stone, and Sarah Worth), as well as the philosophy faculty at my alma mater, the University of South Carolina, and my esteemed colleagues at Greenville Technical College, in the Humanities Department and beyond. Discussing ultimate issues with the students at Greenville Tech constantly reminds me of the excitement and importance of philosophy.

Part of chapter 2 and significant portions of chapter 5 of this book were published as "Hume's Purely Practical Response to Philosophical Skepticism" in *Hume Studies*, 43(2) (November 2017), 3–28, doi:10.1353/hms.2017.0005. Much of chapter 4 appeared as "Hume on the Defeasible Justification of the Vulgar Belief in Body," in *History of Philosophy Quarterly* 36.4 (October 2019), 359–376, https://www.jstor.org/stable/48573190. In both cases, the material has been reworked in various ways. I am grateful to the Hume Society and to the University of Illinois Press for permission to reprint this work. All quotations of Hume's *Treatise*, Abstract, and first *Enquiry* are from the edited versions of these works at *Hume Texts Online* (https://davidhume.org). This free, open-access site is developed and maintained by Amyas Merivale, and the texts are edited by Merivale and Peter Millican. The searchable texts on the site have been invaluable to my research, and I am glad for the chance to thank Merivale and Millican for their contribution to worldwide Hume scholarship.

My most profound thanks are to my wife Patricia—the true, the good, and the beautiful—who has made innumerable sacrifices to support my philosophical pursuits. Wise men proportion their belief to the evidence, says Hume, but Patricia has always believed in me far more than the evidence could justify. She and our daughter Irene are my connections to the external world and to the joy of living. It is to them that I dedicate this book.

List of Abbreviations

Works by David Hume

AT *An Abstract of a Book lately Published; Entituled, A Treatise of Human Nature, &c. Wherein the Chief Argument of that Book is farther Illustrated and Explained*, at *Hume Texts Online* (https://davidhume.org), eds. Amyas Merivale and Peter Millican. Citations indicate paragraph numbers.

EHU *An Enquiry Concerning Human Understanding*, at *Hume Texts Online* (https://davidhume.org), eds. Amyas Merivale and Peter Millican. Citations indicate section and paragraph numbers.

EPM *An Enquiry Concerning the Principles of Morals*, ed. Tom L. Beauchamp (Oxford: Oxford University Press, 1998). Citations indicate section and paragraph numbers.

DNR *Dialogues Concerning Natural Religion and Other Writings*, ed. Dorothy Coleman (Cambridge: Cambridge University Press, 2007). Citations indicate part and paragraph numbers.

HL *Letters of David Hume*, 2 vols., ed. J. Y. T. Greig (New York: Oxford University Press, 1932). Citations indicate volume and page numbers.

L *Letter from a Gentleman to His Friend in Edinburgh*, eds. Ernest C. Mossner and John V. Price (Edinburgh: Edinburgh University Press, 1967). Citations indicate page numbers.

MOL "My Own Life," in *Essays Moral, Political, Literary,* ed. Eugene F. Miller (Indianapolis: Liberty Fund, 1987). Citations indicate page numbers.

NHR *The Natural History of Religion*, in *The Natural History of Religion: A Critical Edition*, ed. Tom L. Beauchamp (Oxford:

Clarendon Press, 2007). Citations indicate section and paragraph numbers.

T *Treatise of Human Nature*, at *Hume Texts Online* (https://davidhume.org), eds. Amyas Merivale and Peter Millican. Citations indicate book, part, section, and paragraph numbers.

Introduction

The Problem of Hume's Skepticism

HUME'S SKEPTICISM VERSUS HIS SCIENCE OF HUMAN NATURE

This book addresses the central interpretive question about the philosophy of David Hume: how does his skepticism relate to his constructive science of human nature? Scholars often frame this issue in terms of the relationship between Hume's skepticism and his naturalism.[1] "Naturalism" is not a term Hume uses, but usually commentators use it to refer to his researches in the science of man. The puzzle, stated in these terms, is that skepticism and naturalism seem incompatible. If skeptics refrain from making claims about how the world is, and scientists are in the business of making claims about how the world is, then it seems as if no skeptic can be a scientist. But Hume considers himself both a skeptic and a scientist.

On one hand, textual evidence for Hume's skepticism abounds. He presents skeptical arguments in sections entitled "Of scepticism with regard to reason" (T 1.4.1), "Of scepticism with regard to the senses" (T 1.4.2), and "Sceptical doubts concerning the understanding" (EHU 4). He also presents catalogues of several skeptical arguments in "Conclusion of this book" (T 1.4.7) and "Of the Academical or Sceptical Philosophy" (EHU 12, Parts 1 and 2). Hume even seems to express personal despair at points in the face of these skeptical arguments (T 1.4.2.56–57, 1.4.7.1–8). Furthermore, he

1. Norman Kemp Smith first framed the issue in terms of skepticism and naturalism in his epochal *The Philosophy of David Hume* (New York: Palgrave Macmillan, 1941, 2005), 129. See the discussion in Don Garrett's "Introduction" to the 2005 reissue, especially xxxiv. The terminology of skepticism and naturalism is not especially illuminating, unless we stipulate what we mean by each of these terms. Each can have a wide range of meanings in contemporary philosophical usage, and "naturalism" is not a term Hume uses at all.

endorses skepticism and self-identifies as a skeptic throughout his corpus. In the final section of Book 1 of the *Treatise*, he says, "In all the incidents of life we ought still to preserve our scepticism. . . . Nay if we are philosophers, it ought only to be upon sceptical principles" (T 1.4.7.11). In his "Abstract" of the *Treatise*, he describes his philosophy as "very sceptical" (AT 27). In *An Enquiry Concerning Human Understanding* he champions "the ACADEMIC or SCEPTICAL philosophy," which he also describes as "*mitigated* scepticism" (EHU 5.1–2, 12.24–26). Philo, widely regarded as the character who speaks for Hume in *Dialogues Concerning Natural Religion*, is also a mitigated skeptic (DNR Part 1).[2] Hume is commonly credited with exposing the "problem of induction" in his analysis of causal inference, a problem that, in the eyes of many philosophers, raises an important skeptical challenge to a wide swath of ordinary beliefs. No wonder that, as D. C. Stove remarks, Hume has so often been classified as "the breaker *par excellence*," "pre-eminent as a skeptical or critical philosopher: one whose *forte* consists in casting doubt on accepted beliefs by exposing the weakness of accepted inferences."[3] Arguably, Hume's destructive powers are chiefly what have earned him a place in the philosophical pantheon.

On the other hand, even a casual survey of the titles of Hume's works shows that his primary goal is not to tear down, but to build up, the edifice of human beliefs. The title of *A Treatise of Human Nature* indicates that Hume's aims are constructive, descriptive, and scientific, and the subtitle (*being an attempt to introduce the experimental method of reasoning into moral subjects*) substantiates this as well.[4] He announces his project as "the science of man," a science itself founded on "experience and observation" (T Intro. 5–6). He stands on the shoulders of Bacon, Locke, Shaftesbury, Mandeville, Hutcheson, and Butler (T Intro. 7). The work divides into three books, each of which constructively describes the understanding, the passions, and morals, respectively. Only the fourth part of the first book ("Of the Understanding") deals with "the sceptical and other systems of philosophy." Furthermore, Hume's discussion of skepticism does not end his project: he goes on for another two books to constructively describe passions and morality. Hume recasts the three books of the *Treatise* in three stand-alone works: *An Enquiry Concerning Human Understanding*, *A Dissertation on the Passions*, and *An Enquiry Concerning the Principles of Morals*, each of which has the positive

2. Philo expresses a position that verbally echoes and materially coincides with what Hume describes as "mitigated scepticism" in Section 12 of EHU. See especially Philo's speech in DNR 1.8–12. Philo neither asserts nor denies explicitly that he is a skeptic. The other characters assert that he is one. Philo responds by clarifying his actual position.

3. D. C. Stove, "Hume, Kemp Smith, and Carnap," *Australasian Journal of Philosophy* 55, no. 3 (December 1977): 189.

4. In this context, by "moral subjects," Hume does not mean "ethics" but "mental subjects." Elsewhere he defines "Moral philosophy" as "the science of human nature" (EHU 1.1).

descriptive thrust of the *Treatise*. (The first *Enquiry* treats skepticism proportionally about as much as does Book 1 of the *Treatise*). In Hume's *Essays: Moral, Political, and Literary*, he develops his views of political science, economics, aesthetic criticism, and other topics. His six-volume *History of England* is a work of constructive historical research. *The Natural History of Religion*, as its title indicates, provides a naturalistic account of the origins of religious belief. Perhaps the only one of Hume's works that is almost entirely negative, rejecting more claims than it establishes, is his posthumously published *Dialogues Concerning Natural Religion*. But Hume's target in that work is in a unique and restricted domain.

I propose to frame the discussion of Hume's skepticism and naturalism around a more specific question: "Is Hume an epistemic skeptic about core beliefs?" The terms "epistemic," "skeptic," and "core belief" need some unpacking. First, by "core belief" I intend to designate a specific domain of beliefs about which Hume might be skeptical. Skepticism can apply to any given domain of beliefs.[5] The importance of skepticism varies depending on the domain of belief toward which it is directed. Skepticism about the number of blades of grass on my lawn is trivial. Skepticism about the existence of God may be of more significance. Perhaps the most important domain to which skepticism might apply is the domain of what I call "core beliefs." By "core belief," I mean a belief that is practically indispensable both for ordinary life and for scientific research. On my account, Hume's skepticism chiefly targets two core classes of beliefs: beliefs produced by reason and sensory belief—that is, belief in bodies that arises from sensation.[6] While we might be able to get along in life or science without some particular members of these classes of beliefs, we cannot get along without the classes as wholes. These two classes of beliefs are collectively, though not distributively, core classes of beliefs.[7] When I refer to these "core beliefs," I mean these two classes of beliefs, not simply individual members of them. Hume presents many arguments against beliefs that his contemporaries commonly accepted—for example, belief in substances in which accidents inhere

5. Don Garrett, "Hume's Conclusions in 'Conclusion of this Book,'" in *The Blackwell Guide to Hume's* Treatise, edited by Saul Traiger (Malden, MA: Blackwell, 2006), 171; *Hume* (New York: Routledge, 2015), 215.

6. As I explain in chapter 1, Hume clarifies that, strictly speaking, the senses do not produce belief in continued and distinct existences; principles of the imagination produce these beliefs (T 1.4.2.3–14). However, in the title of T 1.4.2 and elsewhere, he uses "the senses" more broadly to refer to the processes that produce beliefs in continued and distinct existences. It is this broader usage that I intend when I say that beliefs produced by the senses are, as a class, core beliefs. According to Peter Millican and Hsueh Qu, belief in continued and distinct existences is dispensable for Humean science. Millican, "Hume's Chief Argument," in *The Oxford Handbook of Hume*, edited by Paul Russell (New York: Oxford University Press, 2016), 99, 100; Qu, *Hume's Epistemological Evolution*, 192–93. I reply to this position in chapter 7.

7. Thanks to Don Garrett for helping me clarify this point.

(T 1.1.5, 1.4.3). However, he does not treat this belief as at all useful for common life or science. Arguments against dispensable beliefs are perhaps better characterized as critical rather than skeptical. Arguments against core beliefs comprise the heart of Hume's skeptical crisis.

Skepticism about a domain of belief can come in different varieties.[8] In general, the skeptic points out some kind of defect or flaw in a belief. For example, a practical skeptic may claim that holding some belief violates prudential or ethical norms—regardless of whether the belief has adequate grounds or warrant.[9] Another kind of skeptic might claim that some belief is psychologically impossible to hold or give up. A truth skeptic may claim that some belief is simply false. An epistemic skeptic denies that a belief is epistemologically justified.[10] My main question about Hume is whether he is an epistemic skeptic, and about which beliefs, and why. However, defining skepticism in terms of epistemic justification may seem to trade one terminological ambiguity for another. We need to define epistemic justification in order to know what the epistemic skeptic denies of our beliefs.

At a first approximation, epistemic justification has to do with the intellectual respectability of our beliefs.[11] We can get an initial grasp of the concept by way of contrast: beliefs formed by way of wild guesses or wishful thinking are not epistemologically justified.[12] We have grounds on the basis of which we hold justified beliefs; ungrounded beliefs are unjustified.[13] As Robert Audi says, epistemically justified beliefs "are quite in order from the point of view of the standards for what we may reasonably believe."[14] Put another way, holding an epistemologically unjustified belief violates epistemic

8. See for example the taxonomies of skepticism given by Robert J. Fogelin and Don Garrett. Robert J. Fogelin, *Hume's Skepticism in the Treatise of Human Nature* (Boston: Routledge and Kegan Paul, 1985): 5–6; Don Garrett, "Hume's Conclusions," 170–71, and *Hume*, 215–18.

9. This seems to be what Fogelin means by "prescriptive skepticism" or, in a later work, "belief skepticism." *Hume's Skepticism*, 5; Robert J. Fogelin, *Hume's Skeptical Crisis* (New York: Oxford University Press, 2009): 49. Cf. Garrett, "Hume's Conclusions," 171.

10. This seems to be what Fogelin means by "theoretical skepticism" and what Garrett means by "epistemic merit skepticism." Fogelin, *Hume's Skepticism*, 5; Garrett, "Hume's Conclusions," 171. Fogelin uses the term "epistemological skepticism" to refer to skepticism about beliefs that are intelligible but lack adequate grounds. He contrasts it with conceptual skepticism, which denies the intelligibility of a claim. I do not have this distinction in view when I talk about "epistemic skepticism."

11. There is no uncontroversial definition of the concept among contemporary epistemologists. William P. Alston, *Beyond "Justification:" Dimensions of Epistemic Evaluation* (Ithaca, NY: Cornell University Press, 2005), 11–28. Alston argues that we should dispense with the notion: "the perennial quest for what it is for a belief to be justified, and what are the necessary and/or sufficient conditions for such a status, is quixotic, of the same order as the search for the Fountain of Youth." Ibid., 11.

12. Robert Audi, *Epistemology: A Contemporary Introduction to the Theory of Knowledge*, 2nd ed. (New York: Routledge, 2003): 2.

13. Ibid., 2–3.

14. Ibid., 2. Among Hume scholars, Kevin Meeker assumes that Hume is concerned with knowledge, understood as justified true belief. Meeker, "Hume: Radical Skeptic or Naturalized Epistemologist?" *Hume Studies* 24, no. 1 (April 1998): 34.

norms.[15] Epistemologically justified beliefs are those that we ought to hold, or at least permissibly hold, from an epistemic perspective. Epistemologically unjustified beliefs are beliefs we ought not to hold, from the epistemic perspective.

The exact nature of epistemic skepticism varies depending upon the skeptic's preferred account of epistemic justification and its norms. But generally speaking, the epistemic skeptic denies that a targeted belief is the sort we ought to hold, from the epistemic perspective. It is not the sort that typically turns out to be true; it lacks a feature which is necessary for knowledge. According to the epistemic skeptic, the targeted beliefs bear more of a family resemblance to guesses than to intellectually respectable beliefs.

These terminological glosses suffice to explain the general question this book addresses.[16] When I ask, "Is Hume an epistemic skeptic about core beliefs?" I mean, are there any classes of beliefs which Hume thinks are practically indispensable to life and science, but which he also thinks are groundless, unwarranted, intellectually disreputable? Are there any beliefs that are practically necessary for life and science, but which we ought not to hold, from that perspective which aims at truth? In particular, does Hume regard the beliefs produced by reason and sensory beliefs as epistemologically unjustified?

SKEPTICAL INTERPRETATIONS

A historic tradition of interpretation sees Hume as an epistemic skeptic about core beliefs. Hume's major interpreters from the eighteenth through the nineteenth century read him this way.[17] According to Thomas Reid, Hume "built a system of scepticism, which leaves no ground to believe any one thing rather than its contrary."[18] Immanuel Kant, for all his respect for

15. Some readers may wonder whether talk of epistemic norms presupposes direct voluntary control over belief formation, on the assumption that "ought implies can." I do not mean to presuppose any substantive commitments about the nature of or preconditions for normativity. I use this language in the roughly the same sense that we might say "You shouldn't believe everything you hear." We mean that those who believe everything they hear fall short of an epistemic standard, whether or not they have the ability to do otherwise.

16. I discuss each of these concepts in more detail in future chapters. For example, I have passed over Hume's distinctions between "antecedent" and "consequent" skepticism, mitigated skepticism and excessive (EHU 12), because they are peripheral to clarifying my overarching question, but I take them up in chapter 7.

17. For a review of this early tradition of skeptical interpretation, see Kemp Smith, *The Philosophy of David Hume*, 3–8, 79–88; Garrett, "Introduction," xxvi–xxvii; Janet Broughton, "Hume's Naturalism and His Skepticism," in *A Companion to Hume*, ed. Elizabeth S. Radcliffe (Malden, MA: Blackwell, 2008), 425–26.

18. Thomas Reid, *The Works of Thomas Reid*, vol. 1, *Inquiry into the Human Mind*, 7th ed., edited by William Hamilton (Edinburgh: Maclachlan and Stewart, 1872): 95b.

Hume, concludes that he "deposited his ship on the beach (of skepticism) for safekeeping, where it could then lie and rot."[19] In the late nineteenth century Thomas Hill Green provided a similar narrative of Hume's place in the history of philosophy.[20] Hume took Locke's principles to their logical skeptical conclusion, thus raising the question that only Kant's philosophy could adequately answer.[21] Although Reid, Kant, and Green give different analyses of the origin of Hume's skepticism, they all agree that he is an epistemic skeptic about core beliefs.

The skeptical interpretive tradition remains alive today. Contemporary epistemologists still regularly depict Hume as an inductive skeptic.[22] Many Hume specialists in the twentieth and twenty-first centuries also continue to read Hume as an epistemic skeptic.[23] Although they often offer widely varying accounts of Hume's arguments at the level of detail, they see him as denying that one or more of our core beliefs are epistemologically justified.

The great advantage of this time-honored reading of Hume is that it easily makes sense out of what he says about skepticism. On this view, he calls himself a skeptic because he acknowledges the epistemic force of the skeptical arguments that he puts forward. These arguments do not target a few trivial or dispensable beliefs in a restricted domain, but a broad range of highly significant beliefs. Thus, his philosophy is "very skeptical."

But the skeptical reading of Hume faces a great difficulty: making sense out of his positive scientific project.[24] First, how does Hume's constructive

---

19. Immanuel Kant, *Prolegomena to Any Future Metaphysics*, trans. Gary Hatfield, in *Theoretical Philosophy after 1781*, ed. Henry Allison and Peter Heath (New York: Cambridge University Press, 2002): 4:262; 58.

20. T. H. Green, *Hume and Locke* (New York: Thomas Y. Crowell Co., 1968).

21. Ibid., 2–3.

22. See for example Alston, *"Justification,"* 215, 221–24, 233; Audi, 296–98 and 310–13; John Greco, "Virtues in Epistemology," *The Oxford Handbook of Epistemology*, ed. Paul K. Moser (Oxford: Oxford University Press, 2002): 304–5; Alvin Plantinga, *Warrant and Proper Function* (Oxford: Oxford University Press, 1993): 124–28.

23. Barry Stroud, *Hume* (New York: Routledge and Kegan Paul, 1977); Fogelin, *Hume's Skepticism*; Louis E. Loeb, *Stability and Justification in Hume's Treatise* (New York: Oxford University Press, 2002); Meeker, "Hume: Radical Sceptic"; cf. Meeker's *Hume's Radical Scepticism and the Fate of Naturalized Epistemology* (New York: Palgrave Macmillan, 2013); H. O. Mounce, *Hume's Naturalism* (New York: Routledge, 1999): 53–61; Harold W. Noonan, *Hume on Knowledge* (New York: Routledge, 1999): especially 12–14, 45–46, 129–30, 161–62; Ira Singer, "Nature Breaks Down: Hume's Problematic Naturalism in *Treatise* I iv," *Hume Studies* 26, no. 2 (November 2000): 225–43; cf. Ira Singer, "Hume's extreme skepticism in *Treatise* I IV 7," *Canadian Journal of Philosophy*, 25, no. 4 (December 1995): 595–622; Broughton, "Hume's Naturalism," 425–40; cf. Janet Broughton, "The Inquiry in the *Treatise*," *Philosophical Review* 113, no. 4 (November 2004): 537–56; Karann Durland, "Extreme Skepticism and Commitment in the *Treatise*," *Hume Studies* 37, no. 1 (April 2011): 65–98; Graciela De Pierris, *Ideas, Evidence, and Method: Hume's Skepticism and Naturalism Concerning Knowledge and Causation* (New York: Oxford University Press, 2015).

24. Donald C. Ainslie, following Philip Cummins, calls this the "integration problem." Ainslie, *Hume's True Scepticism* (New York: Oxford University Press, 2015), 35, 221, 227–30, citing Cummins, "Hume's Diffident Skepticism," *Hume Studies* 25, no. 1/2 (April/November 1999): 43–65.

goal of describing human nature relate to his negative goal of showing that our core beliefs are epistemologically unjustified? This is a question about what Hume is trying to do. Second, does Hume provide any principled defense of moving on with a scientific project once he has exposed our core beliefs as epistemologically unjustified? If the first question is about what Hume is trying to do, then the second question is about how he is trying to do it.[25]

Skeptical interpreters have proposed three main approaches to the first question, "How does Hume's positive scientific goal of describing human nature relate to his negative goal of showing that our core beliefs are epistemologically unjustified?" First, some have simply denied that Hume ever really has a positive scientific project. His aims are only skeptical and destructive all along. Reid, for example, writes

> It seems to be a peculiar strain of humour in this author, to set out in his introduction, by promising, with a grave face, no less than a complete system of the sciences, upon a foundation entirely new, to wit, that of human nature; when the intention of the whole work is to show, that there is neither human nature nor science in the world.[26]

Reid's suggestion—that Hume's positive project is literally a joke—eliminates the need to answer the second question, as to whether Hume provides a principled defense of continuing with science in the face of epistemic skepticism. However, the complete denial that Hume has a positive project is simply incredible and has no modern adherents.

Second, some skeptical interpreters propose that Hume does start out with a positive scientific project, but his skepticism sabotages it.[27] Once Hume has denied that our core beliefs are epistemologically justified, he has in effect denied that any of his descriptive claims merit acceptance. Although Hume starts out with the sincere intention to develop a well-grounded science of

---

25. Paul Russell helpfully distinguishes between kinds of two solutions to the Kemp Smith problem. An interpretive solution explains what Hume intends to do with skepticism and naturalism. A philosophical solution shows that what he is doing is coherent and credible. Russell, *The Riddle of Hume's* Treatise*: Skepticism, Naturalism, and Irreligion* (New York: Oxford University Press, 2008), 7–8.

26. Reid, *Inquiry into the Human Mind*, chapter 1, Section 5, 102a.

27. Janet Broughton, "The Inquiry," 550–53. Singer argues that Hume's extreme skepticism gets out of his control undermines his positive project in spite of Hume's intentions. Singer, "Nature Breaks Down," 236. Loeb thinks that Hume starts with a constructive epistemological project which ends in negative skepticism. Loeb, *Stability*, viii, 12, 16. Against Singer, Loeb claims that Hume does not arrive at the skeptical nadir unwillingly or in spite of his intentions. Peter Millican thinks that Hume has no good response to skepticism in the *Treatise*. He specifically rejects the Title Principle (discussed below) as an effective or important Humean solution to skepticism. Millican, "Hume's Chief Argument," 97–98, 105n42; Millican, "Skepticism about Garrett's *Hume*: Faculties, Concepts, and Imposed Coherence," in *Hume Studies* 40(2) (2014), 222n16.

man, he finds to his own chagrin that the project founders on skeptical problems. His works are the candid records of a thwarted ambition.

The weakness of the skepticism-thwarts-science reading is its inability to answer the second question: what principled defense does Hume give for continuing with science in the face of skepticism? It is sometimes suggested that Hume keeps going with science because he cannot help it. Many commentators point out Hume's insistence that our natural core beliefs simply will not go away permanently; they are psychologically compulsory, and we resume them, willy-nilly, even in the face of the skeptical dilemma.[28] But it is not clear how psychological irresistibility legitimizes epistemologically unjustified core beliefs.[29] Hume says that beliefs inculcated by mere repetition from infancy are sometimes psychologically ineradicable (T 1.3.9.18). But he regards such beliefs as epistemologically unjustified and, often, practically deleterious. So irresistibility alone will not legitimize Hume's core beliefs.[30] Adherents of the skepticism-thwarts-science position are committed to the view that Hume carries on with science and common life without having any viable principled reason for doing so.[31] But this is deeply unsatisfying. It seems unlikely that a philosopher of Hume's stature would leave such a massive problem running through his entire corpus without addressing it somehow.

A third kind of skeptical interpretation proposes that Hume does have a scientific project, and that he uses skepticism as a tool to support and motivate

28. For example, Stroud, *Hume*, 237–38; Fogelin, "Hume's Skepticism," in *The Cambridge Companion to Hume*, 2nd ed., edited by David Fate Norton and Jacqueline Taylor (Cambridge: Cambridge University Press, 2009), 234; Noonan, *Hume on Knowledge*, 46; Broughton, "Hume's Naturalism," 435.

29. In chapter 1, I argue that irresistibility plays a role in the epistemic justification of beliefs, but the mere fact that a belief is irresistible does not make it justified overall or invulnerable to defeaters.

30. Many commentators make this point. See Stroud, *Hume*, 247–49; William Edward Morris, "Hume's Conclusion," *Philosophical Studies* 99 (Netherlands: Kluwer, 2000), 94; Michael Ridge, "Epistemology Moralized: David Hume's Practical Epistemology," *Hume Studies* 29(2), November 2003, 171–75; Meeker, "Hume: Radical Skeptic," 37. Broughton, "Hume's Naturalism," 435–36. Fogelin seems to have increasingly lost confidence that Hume has a rationale for continuing with his scientific project. In Fogelin's earlier work, he argues that after his skeptical crisis, Hume "can then turn, *with perfect justification*, to the factual question of how humans are able to form beliefs despite the skeptical arguments that can be brought against them" (emphasis mine). Fogelin, *Hume's Skepticism*, 146. But in his later work, Fogelin seems less sanguine. Hume simply blindly submits to the natural beliefs of sense and understanding, pursuing philosophy when it seems fun to do so, but without any satisfying defense of it. *Hume's Skeptical Crisis*, 125–37.

31. Many commentators have expressed doubt about Hume's ability to provide a principled rationale for carrying on in the face of skepticism. Reid sees Hume's "absolute scepticism" as "destructive of . . . the science of a philosopher, and of the prudence of a man of common understanding." Reid, *Inquiry into the Human Mind*, 95b. For Kant (and Green after him), only idealism can save science from Humean skepticism. Noonan says, "It is not evident that these questions have any complete answers." Noonan, *Hume on Knowledge*, 46. Durland argues more forcefully that no one has yet, nor is likely to ever, propose a satisfying account of how he might overcome his extreme skepticism. Durland, "Extreme Skepticism." In different ways, Meeker and Singer both argue that even a naturalized epistemology does not save Hume from skepticism or give him a compelling rationale for moving on from it. Meeker, "Hume: Radical Sceptic"; Singer, "Nature Breaks Down."

that project.[32] Hume's skepticism "clears the ground" for his naturalistic description of human nature. Once we see that all of our beliefs fall hopelessly short of rational epistemic standards, we are free to investigate the natural causal processes that produce those unjustified beliefs. The suggestion here seems to be along the lines of W. V. O. Quine's proposal in "Epistemology Naturalized" that we quit the hopeless task of trying to provide epistemic foundations for science, and instead just proceed with science and see what it tells us about human nature.[33] Graciela De Pierris defends a variant of this interpretation. In De Pierris's view, Hume's radical skeptical reflections presuppose different normative requirements than those of science and ordinary life.[34] Normal science is insulated from radical skepticism: "even if we reach a negative (skeptical) conclusion at the second-order or meta-level, this does not *by itself* imply that we must also reject the normative force of the methods employed in what we take to be our best first-order inquiries."[35] On De Pierris's view, radical skepticism clears the ground for science in the sense that it guards natural philosophers in the Newtonian tradition from drawing metaphysical or theological conclusions from their science.[36]

This reading faces two problems. First, it relies on the dubious claim that Hume distinguishes epistemic from scientific normativity.[37] If it turns out (as I hold) that Hume does not make this distinction, then he cannot insulate ordinary science from epistemic skepticism. Second, even if he distinguishes epistemology from science and dispenses with the former, Hume needs to provide a reason to prefer science and its deliverances to other live methodological alternatives, such as "superstition." If Hume gives up the notion that his scientific account of human nature is epistemologically superior to religious accounts, then it is not clear why his account is preferable at all. In other words, Hume still has not given a principled defense of carrying on with science (rather than alternative methods of inquiry) in the face of skepticism.

---

32. Stroud, *Hume*, 1–16; Fogelin, *Hume's Skepticism*, 146.

33. W. V. Quine, "Epistemology Naturalized," in *Ontological Relativity and Other Essays* (New York: Columbia University Press, 1969): 69–90.

34. De Pierris, *Ideas*, 20–21.

35. Ibid., 21. There is an interesting convergence between De Pierris's radical skeptical interpretation of Hume, and the view of Ainslie, who rejects skeptical readings. According to Ainslie too, Hume's skepticism is restricted to a second-order level of inquiry and does not impinge on first-order scientific inquiry. Ainslie, *Hume's True Scepticism*, 243–45. Ainslie concludes that Hume is *not* a total skeptic because his skepticism is restricted to the meta-level; De Pierris concludes that Hume *is* a radical skeptic because he is a skeptic at the meta-level. When I talk about epistemic skepticism, I mean skepticism about first-order beliefs, regardless of what is happening at the meta-level. Understood in these terms, neither De Pierris nor Ainslie has a skeptical reading of Hume.

36. De Pierris, *Ideas*, 23, 296–306.

37. Alston notes that many of the pre-nineteenth philosophical works that we think of as making a contribution to epistemology, are actually "treatments of methods of intellectual inquiry or of the logic of science." Alston, *Beyond "Justification,"* 2. The idea of a clear distinction between scientific method and epistemology seems to be of recent vintage.

In summary, the great objection to skeptical interpretations of Hume is rooted in a philosophical objection. The philosophical objection is that if core beliefs are epistemologically unjustified, then there is no rationale for continuing with science. This objection applies to epistemic skepticism in general, but it also applies to the skeptical interpretation of Hume in particular. If we charitably regard Hume as a great philosopher, then we should assume that he will not adopt incoherent positions. It is incoherent to pursue science as an epistemic skeptic. Therefore, if Hume is a great philosopher, then we should assume he is not an epistemic skeptic, unless there is overwhelming textual evidence that he has made this catastrophic blunder.[38]

NON-SKEPTICAL INTERPRETATIONS

Perhaps because of the difficulties that face the skeptical interpretation of Hume, since the early twentieth century many commentators have understood him as a scientist who is not ultimately an epistemic skeptic at all. Everyone must admit that Hume is some kind of skeptic, but we can call this the "non-skeptical tradition" in the sense that it denies that Hume is an epistemic skeptic about core beliefs.

Norman Kemp Smith stands at the head of this non-skeptical tradition.[39] Kemp Smith takes Hume's statement that "Reason is, and ought only to be the slave of the passions" (T 2.3.3.4) as a programmatic statement not only for Hume's theory of motivation and morality, but also for his epistemology.[40] As applied to Hume's epistemology, the maxim means that reason not only is, but epistemologically ought to be, the slave of the natural beliefs.[41] So our natural core beliefs are justified, regardless of whether reason supports or even opposes them. In short, Hume's key move is to adopt a permissive epistemology, one in which instinct rather than argument is authoritative, and in which some instinctive beliefs that fall afoul of reason are epistemologically justified. Since natural core beliefs (whatever their relation to reason) retain their positive epistemic status, so too does Hume's entire scientific

38. Hsueh Qu makes this argument especially clearly in *Hume's Epistemic Evolution*, 132–33, 209.

39. See his "The Naturalism of Hume (I)," in *Mind*, 14, no. 54 (April 1905), 149–73 as well as his further elaboration of his position in *The Philosophy of David Hume*. Garrett's "Introduction" to *The Philosophy of David Hume*, and Loeb, *Stability*, 20–21, provide helpful summaries of the Kemp Smith reading.

40. Kemp Smith argues that Hume developed his theory of the passions and morality prior to developing his epistemology, and that he wrote Books 2 and 3 of the *Treatise* before writing Book 1. *The Philosophy of David Hume*, Preface and chapter 1. Book 1 applies the reason/feeling maxim to epistemology.

41. Ibid., 83–87.

project. Hume's "skepticism" consists merely in restricting our enquiries to the domain of experience, distrusting our capacity for artificial speculation, and combating our credulous tendencies.[42]

In general, recent non-skeptical interpretations follow the lead of Kemp Smith by attributing to Hume a permissive epistemology that saves the positive epistemic status of core beliefs which might appear to be threatened by skeptical arguments. More specifically, recent interpreters deal with Hume's skeptical arguments in one of two ways. First, some readers think that Hume's skeptical arguments do attack the epistemic justification of their targeted beliefs, but that Hume himself does not endorse the presuppositions of these skeptical arguments.[43] We may refer to these as *reductio* readings, since they construe Hume's skeptical arguments as reductio ad absurdum arguments against assumptions he ultimately rejects.[44] Second, some readers hold that Hume's skeptical arguments are not even intended to discredit the epistemic justification of their targeted beliefs.[45] They simply describe the loss of confidence that does or can occur when we reflect in certain ways; the arguments do not entail any epistemic evaluation of that loss of confidence. So Hume has no epistemic skepticism to overcome. Some non-skeptical interpreters do not fit neatly into these two categories, but nevertheless hold that Hume's final epistemic position neutralizes the skeptical threats he may face.[46]

---

42. Ibid., 132

43. David Owen points out that these readings do construe the skeptical arguments as epistemic. Owen, *Hume's Reason* (New York: Oxford University Press, 1999), 198.

44. Annette C. Baier, an influential proponent of this kind of interpretation, says Book 1 of the *Treatise* constitutes a "*reductio ad absurdum* of Cartesian intellect." Baier, *A Progress of Sentiments: Reflections on Hume's* Treatise (Cambridge, MA: Harvard University Press, 1991): 21. Morris defends a version of this view in "Hume's Conclusion." Although Michael P. Lynch argues against the details of Morris's interpretation of *Treatise* 1.4.1, he also holds something like a *reductio* view. Lynch, "Hume and the Limits of Reason," *Hume Studies* 22, no. 1 (April 1996): 89–104. Lynch's Hume is running a kind of reductio ad absurdum argument against those who think they can evade epistemic circularity. Ridge seems to think that Hume's skeptical arguments first drive him to despair, and then drive him to change his epistemology by adopting the Title Principle as an epistemic norm which neutralizes the threats. "Epistemology Moralized," especially 177–78, 196–98n8, 189. According to Ainslie, total skepticism is generated by a false philosophy that demands proof of the reliability of our fundamental cognitive tendencies. Ainslie, *Hume's True Scepticism*, 240–41. *Treatise* 1.4.7 narrates Hume's own transition from false to true philosophy. Ibid., 238. For Ainslie and Ridge, the reductio ad absurdum argument is one which applies to Hume's own initial position, driving him to adopt another position.

45. See especially Garrett, *Cognition and Commitment in Hume's Philosophy* (New York: Oxford University Press, 1997), chapter 10, 205–42; Garrett, *Hume*, 213–44; Owen, *Hume's Reason*, chapters 8–9, 175–223.

46. Frederick F. Schmitt splits the difference between skeptical and non-skeptical readings of the *Treatise* by arguing that, of the two core beliefs on which Hume focuses (beliefs produced by inductive inferences and belief in bodies), the former is epistemologically justified and the latter is not. Schmitt, *Hume's Epistemology in the* Treatise*: A Veritistic Interpretation* (New York: Oxford University Press, 2014), 368–75. Mounce also thinks that Hume escapes the force of the skeptical argument against reason (T 1.4.1) but succumbs to the force of the skeptical argument against belief in body (T 1.4.2). *Hume's Naturalism*, 49–61.

Many non-skeptical readings crucially invoke Hume's so-called Title Principle as his final epistemic norm which allows him to reject or overcome epistemic skepticism in the *Treatise*.[47] The Title Principle says

> Where reason is lively, and mixes itself with some propensity, it ought to be assented to. Where it does not, it never can have any title to operate upon us. (T 1.4.7.11)

If the normative force of the "ought" (and "title") in the Title Principle is epistemic, then Hume intends it to describe which propositional attitudes toward the deliverances of reason are epistemologically justified, and which are not. Lively deliverances of reason "ought to be assented to;" doubt and denial are epistemologically unjustified attitudes. On the other hand, we are epistemologically justified in doubting or even denying non-lively deliverances of reason, to which we have no propensity to assent. The Title Principle neutralizes the skeptical argument against reason (T 1.4.1), which is the argument that in turn generates a "dangerous dilemma" at the skeptical nadir of the *Treatise* (T 1.4.7.7). As Hume points out, we have no propensity to follow reason down the long trail to recursive self-defeat which the argument against reason involves. A trivial quality of the imagination psychologically prevents us from putting much confidence in very refined conclusions such as these, no matter how sound the argument may be (T 1.4.7.6–7). So the Title Principle permits us to ignore the skeptical argument against reason, and therefore to evade the dangerous dilemma too. The Title Principle therefore rescues Hume from this threat of skepticism.

Non-skeptical interpretations have clear strengths. Most importantly, they easily account for Hume's pursuit of science, since on this reading it faces no epistemic challenge. Furthermore, non-skeptical interpreters have rightly drawn attention to the pivotal role that the Title Principle plays in the conclusion of Book 1 of the *Treatise*. They have made a strong case that Hume does not intend his famous analysis of causal inference as a skeptical argument, at least when he first introduces it in *Treatise* 1.3.6 (as I discuss in chapter 2).

47. Kemp Smith, *The Philosophy of David Hume*, 131; Baier, *Progress of Sentiments*, 280; Garrett, *Cognition*, 234–35; Garrett, *Hume*, 227–37; Owen, *Hume's Reason*, 217; cf. 203n12; Morris, "Hume's Conclusion," 109; Ridge, "Epistemology Moralized," 89; Schmitt, 368–75. Garrett is the first to have dubbed these lines "the Title Principle." Ainslie is the exceptional non-skeptical interpreter who denies that the Title Principle is an epistemic norm. Ainslie, *Hume's True Scepticism*, 233, 243–44. As noted above, he confines Hume's skepticism to the question of the ultimate justification of our cognitive tendencies and thinks the ordinary beliefs and practices of science are insulated from this skepticism. Qu agrees that Hume intends that the Title Principle solve his skeptical crisis. Qu, *Hume's Epistemological Evolution*, chapter 6. However, Qu thinks the Title Principle is actually a failure, as Hume, arguably, later came to recognize. Qu's reading is non-skeptical to the extent that Qu's Hume purports, in the *Treatise*, to have a satisfactory epistemic solution to his skeptical problems.

But non-skeptical readings are not entirely satisfying. For one thing, they have to provide an attenuated account of how Hume's final position is still in some sense skeptical.[48] For example, Annette C. Baier says that Humean skepticism—what he calls "true scepticism"—collapses into mere open-mindedness, undogmatic diffidence, and fallibilism.[49] For Don Garrett and Frederick Schmitt, Hume's philosophy is skeptical because it requires us to doubt our skeptical doubts.[50] For David Owen, Hume's philosophy is "skeptical" in the sense that he does not rely on the isolated faculty of reason.[51] For Donald C. Ainslie, Hume is a "true sceptic" in that he admits that there is no ultimate proof of the reliability of our cognitive faculties.[52] But for none of these readers does Humean skepticism involve a denial of the epistemic justification of core beliefs.

It is not at all clear that an attenuated sort of "skepticism" really does justice to the importance that Hume seems to assign to his sort of skepticism. Any Baconian could espouse open-minded fallibilism. Dogmatists are only too willing to doubt their doubts and keep on believing, come what may. The superstitious do not rely on the isolated faculty of reason. And if the epistemic status of ordinary scientific beliefs is insulated from higher-order skepticism anyway, as Ainslie and De Pierris have it, it is hard to see why Hume would care very much about higher-order skepticism; it presents no live threat to the authority of science. Humean skepticism seems to be stronger and different than the forms that non-skeptical accounts can allow.

Moreover, non-skeptical readings have a hard time explaining why Hume stresses the practical reasons we have for resuming core beliefs after skeptical attacks. If Hume has a purely epistemic solution to skeptical arguments, then these practical reasons are at best incidental benefits. But *Treatise* 1.4.7.9–15 and *Enquiry* 12.21–23 foreground practical considerations.[53] *Enquiry* 12 does not provide any direct, explicit answers to the particular skeptical arguments it catalogues, and the Title Principle makes no appearance. If Hume has an epistemic solution to the skeptical arguments he lists in the *Enquiry*, it is far from obvious what that is.

---

48. Ainslie rightly poses this challenge to Baier's *reductio* reading (he calls it a "dialectical interpretation") and others like it. If Hume ultimately has an epistemic solution to the skeptical problems he raises, "Why then does Hume continue to call himself a 'true' sceptic? It seems that Hume should see himself as an *anti*-sceptic if his move to true philosophy means that the 'desponding reflections' are left behind. I will call this interpretive problem the *scepticism* problem: in what sense does Hume remain a sceptic at the end of CtB [*Treatise* 1.4.7, 'Conclusion of this Book']?" Ainslie, *Hume's True Skepticism*, 237.

49. Ibid., 58, 27.

50. Garrett, *Cognition*, 235–37; Schmitt, *Hume's Epistemology*, 372.

51. Owen, *Hume's Reason*, 220–23

52. Ainslie, *Hume's True Skepticism*, 243–45.

53. When I refer to the "*Enquiry*" without qualification, as I do here, I mean the first *Enquiry, An Enquiry Concerning Human Understanding*.

According to one important line of interpretation, the practical reasons for retaining our core beliefs *are* an epistemic solution to the skeptical difficulties. Hume's idea is that beliefs formed in accordance with the Title Principle are epistemologically justified just because they are useful or agreeable to ourselves or others (that is, practically justified).[54] On this practical-epistemic reading, epistemic justification is in some sense dependent upon practical justification. While the practical-epistemic reading makes excellent sense out of much of the textual data, Hsueh M. Qu convincingly argues that it collapses the distinction between epistemic and moral justification—a distinction which is well-grounded in the *Treatise* and in the *Enquiry*.[55] Hume denies that practical reasons can epistemologically justify our core beliefs. So the practical-epistemic response to skepticism is not one that he can make.

Because of the difficulties facing both skeptical and non-skeptical interpretations, the reconciliation of his science and skepticism remains a central question in Hume scholarship. On the skeptical reading, Hume provides a descriptive account of the understanding only to conclude that our core beliefs are epistemologically unjustified. He then continues to write about human nature, the passions, morality, politics, economics, religion, history, and criticism, without giving any account of why we ought to believe him. On the non-skeptical reading, Hume mounts a number of "skeptical" arguments that do not ultimately threaten the epistemic status of any of our core beliefs. He furthermore describes his own philosophy as "very skeptical" even though he thinks that all of his assertions about the world are epistemologically justified. Both readings present us with a rather mystifying philosopher.

---

54. See especially Ridge, "Epistemology Moralized;" Owen, *Hume's Reason,* 216–22; and Karl Schafer, "Curious Virtues in Hume's Epistemology," *Philosophers' Imprint* 14(1), (January 2014): 1–20. Although these authors agree that utility and agreeability are essential to the epistemic justification of belief, their accounts differ in important ways. Moreover, Schafer thinks that Owen and Ridge subscribe to a purely practical reading of Hume such as I defend below, rather than the sort of practical-epistemic reading that he himself advances. Schafer 4n22. If indeed Owen and Ridge do mean to defend a purely practical reading, then this essay is a fresh defense of their position. But as I read them, Owen and Ridge hold a practical-epistemic view. Ridge says that the Title Principle is "the one clearly normative epistemic principle that he lays down at the end of Book 1" and that it "clearly makes appeal to the immediate agreeableness of relying on the understanding." Ridge, "Epistemology Moralized," 189. Owen describes "sceptical reason" as reason functioning in accordance with the Title Principle, "reason embedded in a sensitive nature with properties that allow it to function in the correct way." Owen, *Hume's Reason*, 217. Embedded reason (i.e., the Title Principle) is normative for philosophy. But the question still remains, "not what philosophers should do, but whether we should be philosophers." Owen, *Hume's Reason*, 219. Owen's Hume answers that the pursuit of philosophy is practically justified: it is "positively required for the good life; not just for the pleasure it brings the practitioner, but for the good it can do society." Owen, *Hume's Reason*, 220. So for both Ridge and Owen, as I understand them, the Title Principle is an epistemic norm; for Owen, it is internal to the practice of philosophy. As such it can rescue the epistemic status of our core beliefs from the skeptical arguments brought against them. On a purely practical reading, by contrast, the Title Principle is not a correct epistemic norm; it is not internal to the practice of philosophy. Our core beliefs are epistemologically unjustified at the end of the day, and only practically justified.

55. Qu, "Hume's Practically Epistemic Conclusions?" *Philosophical Studies* (2014) 170, 509–23.

PHILOSOPHICAL SKEPTICISM AND PRACTICAL JUSTIFICATION

In this book, I argue that Hume is an epistemic skeptic about core beliefs but that he gives a purely practical justification for continuing to hold them. He denies that core beliefs—namely, sensory beliefs and the deliverances of reason—are epistemologically justified. However, he continues to hold these epistemologically unjustified beliefs on purely practical grounds. To give them up is not only psychologically impossible but also practically self-destructive. He has a purely practical rationale for carrying on with common life and with scientific research even as a skeptic who, when pressed, admits that his beliefs are unjustified.

I begin by arguing that when Hume wants to talk about the sorts of things that contemporary philosophers discuss under the heading of epistemic justification, he talks about "philosophy."[56] He uses the term "philosophy" to refer to a normative method of inquiry and belief-formation that governs the special sciences. For example, certain kinds of probable inferences are "unphilosophical," and some kinds are "receiv'd by philosophers" (T 1.3.13.1). More broadly, philosophers approve of those belief-forming processes which are permanent, irresistible, and universal (T 1.4.4.1–2). They disapprove of those "trivial" belief-forming processes that lack these features. The norms of philosophy are distinct from prudential and moral norms. An action is prudent if it advances our own long-term self-interest. Morally, Hume approves of qualities which are immediately agreeable or useful to their possessors or to others (T 3.3.1.30; EPM 9.1). Philosophy does not evaluate beliefs or belief-forming processes on the basis of their agreeability or utility, either for oneself or others.

Given this notion of Humean "philosophy," I claim that "philosophical" acceptability is usefully understood as what contemporary philosophers call "epistemic justification." Philosophy's norms, which govern doxastic

---

56. Commentators are divided on the question of how far Hume endorses philosophy. Morris writes that "Hume typically speaks of 'philosophers' and their 'usual' practices, not to identify with them, but to dissociate himself from a generally accepted position with which he disagrees" (95). Loeb divides Book 1 of the *Treatise* into a constructive epistemological phase in Parts 1–3, and a destructive skeptical phase in Part 4. Loeb, *Stability*, 12–20, especially 16–17. He ascribes to Hume an ambivalent relationship with "the philosophers." In the constructive phase of his project, Hume aligns himself with the epistemic commitments of the philosophers. In the destructive phase, he distances himself from the epistemic commitments of the philosophers. Jack C. Lyons seems to assume that "philosophy" is a set of normative epistemic principles which Hume endorses but does not lay much emphasis on it. Lyons, "General Rules and the Justification of Probable Belief in Hume's *Treatise*," in *Hume Studies* 27, no. 2 (November 2001), especially 274n15, 270–71. Garrett says little about the meaning of "philosophy," just that it is comprised of "natural philosophy" and "moral philosophy." Garrett, *Cognition*, 3–7. He does however say that Hume endorses philosophy, especially vis-à-vis "superstition."

practices, particularly in the context of the special sciences, are neither prudential nor moral. Philosophy does not aim at interest-satisfaction but at truth. Most contemporary epistemologists would probably regard our best scientific theories as paradigmatic examples of knowledge. Most would probably agree that if any beliefs are epistemologically justified, the products of the scientific method are. Since Hume's "philosophy" constitutes his scientific method, we should understand it as epistemologically normative. At the very least, "philosophical" norms are the closest thing to epistemic norms that Hume has on offer. On my account of Humean epistemic justification, Hume is an epistemic skeptic about core beliefs if and only if he holds that some of our core beliefs fall short of the standards of "philosophy"—that method of inquiry and belief-formation that governs the special sciences.

In order to determine which arguments (if any) Hume regards as epistemologically skeptical threats, we need to determine in more detail what the standards of philosophy really are, since these standards are also Hume's standards for epistemic justification. I argue that Hume has a propensity-based epistemology, wherein the epistemic status of a belief derives from the properties of the propensity that produces it. I also find in Hume a distinction between two levels of epistemic justification, defeasible justification and overall (or ultimate) justification.[57] A defeasibly justified belief is justified for the time being. An ultimately justified belief is justified all things considered. Defeasibly justified beliefs may or may not prove to be ultimately justified. For Hume, a belief is defeasibly justified if and only if it is produced by a permanent, irresistible, and universal propensity. It is justified overall if and only if it is defeasibly justified and faces no undefeated defeaters. A defeater is an epistemic reason for rejecting a belief. Defeaters come in two varieties, rebutting defeaters and undermining (or undercutting) defeaters.[58] A rebutting defeater for some belief *P* is a reason to believe that not-*P* is true. An undermining defeater for some belief *P* is a reason to believe that *P* is not defeasibly justified. Both sorts of defeaters show up in Hume, although of course not under their contemporary names. With Hume's epistemology in hand, we can sort through the array of potentially skeptical arguments he presents and determine which ones, by his own lights, really impugn the epistemic status of core beliefs.

57. This language is not Hume's and may sound anachronistic. In chapter 2, I present textual evidence to support the claim that he materially endorses these concepts, even though he does not use the terms.

58. Cf. Schmitt, *Hume's Epistemology*, 30. The distinction seems to trace to John Pollock and has become common in epistemology. Thomas Kelly, "Evidence," *The Stanford Encyclopedia of Philosophy* Fall 2014 ed., ed. Edward N. Zalta, accessed November 30, 2015, http://plato.stanford.edu/archives/fall2014/entries/evidence/; cf. John Pollock, *Contemporary Theories of Knowledge*. Towota, NJ: Rowman & Littlefield, 1986.

Hume's true skeptical arguments occur precisely in the sections of the *Treatise* we should expect: "Of scepticism with regard to reason" (T 1.4.1), and "Of scepticism with regard to the senses" (T 1.4.2). Both arguments display a similar pattern. The targeted core beliefs are defeasibly justified, but face defeaters from reason. In *Treatise* 1.4.1, Hume shows that the conclusions of all demonstrative and probable inferences face undermining defeaters from reason itself. In *Treatise* 1.4.2, Hume shows that the vulgar belief in "continued and distinct existences" (enduring, mind-independent objects) faces a rebutting defeater from reason. No epistemic rehabilitation of these beliefs is possible. Hume reprises his main arguments against sensory beliefs in *Enquiry* 12.7–16.

Hume's other seeming skeptical arguments, both in the *Treatise* and in the *Enquiry*, presuppose commitments that conflict with Hume's own positions. Hume does not refute them, even though he has refutations available. Instead, he uses these arguments to motivate his readers to accept the philosophical force of skepticism and avail themselves of his purely practical response. I call these skeptical arguments *reductio* arguments, meaning that they are intended to prove that an opponent's assumptions lead to a skeptical conclusion—the conclusion that certain important beliefs lack epistemic justification.

My reading essentially differs from the sort of *reductio* interpretations I referenced earlier. Other *reductio* interpretations are non-skeptical. On these readings, Hume shows that a certain assumption leads to epistemic skepticism. But his point is that we can avoid epistemic skepticism by giving up the assumption. On my reading, Hume shows that a certain assumption leads epistemic skepticism. But his point is not that we should give up the assumption and thereby avoid epistemic skepticism. His point is rather that we should accept the inescapability of epistemic skepticism. This acceptance prepares the way for his purely practical response to skepticism. Another difference between my reading and other *reductio* interpretations is that on my reading, not all of Hume's skeptical arguments are *reductio* arguments. I will argue that he fully accepts the premises of the skeptical arguments against reason and the senses in the *Treatise*, and the premises of the skeptical argument against the senses in the *Enquiry*.

Hume recommends that when philosophy requires us to suspend our core beliefs, we should ignore the demands of philosophy on practical grounds. This is what the Title Principle itself recommends. I agree with Garrett and others who stress that the Title Principle enables Hume to defuse the key skeptical arguments in the *Treatise*. But I argue that the Title Principle is a merely practical norm. It practically justifies our ignoring the skeptical challenge but does not epistemologically justify our beliefs. The Title Principle, by telling us to sometimes ignore the deliverances of reason, contradicts Hume's epistemic norms, which make no such exceptions to the authority of reason. Many non-skeptical interpreters point out that following the Title Principle is practically motivated.[59] I go further by insisting it is purely practical, not epistemic (philosophical). Furthermore, the Title Principle itself is less important than the general lesson that we should follow philosophy only to the limited extent that we have a practical reason for doing so. Hume repeats that general point in the *Enquiry*, thought he does not repeat the Title Principle in particular.

The purely practical reading of Hume's response to skepticism answers the main objection to skeptical interpretations. The objection is that if Hume is a skeptic, then he has no right to continue to hold beliefs and pursue research. But Hume can concede he has no epistemic entitlement to his beliefs yet still defend a practical entitlement to them. As noted above, practical justification for holding some belief is not sufficient for epistemic justification. But a logical entailment of this claim is that epistemic justification for beliefs is not necessary for their practical justification. For someone like W. K. Clifford, who holds that it is ethically wrong, "always, everywhere, and for anyone, to believe anything upon insufficient evidence," epistemic justification and practical justification are coextensive.[60] But it is conceivable that epistemic and practical justification are not coextensive.[61] It might be practically advantageous for me to believe (against my evidence) that I can beat cancer or leap across a wide crevasse: the mere belief will increase the likelihood that I will in fact beat cancer or make the jump successfully.[62] In fact, a few commentators have suggested that on Hume's final position, our core beliefs are epistemologically unjustified but that we are practically justified in holding

59. Kemp Smith writes that natural beliefs pass "the practical test of human validity." Kemp Smith, "The Naturalism of Hume (I)," 152, 155–56. Garrett points out that following the Title Principle is apparently the best way to satisfy our own desires. Garrett, *Cognition*, 234. Elsewhere he writes that "the disposition to reason in accordance with the Title Principle does indeed achieve moral approval in the *Treatise* as a trait that is useful to its possessor." Garrett, *Hume*, 232–33. Owen also stresses that Hume's Title Principle recommends philosophy and reason ("not functioning in isolation, but embedded in a feeling creature") on practical and moral grounds. Owen, *Hume's Reason*, 217, 211–23. Hume's preference for philosophy and reason "is the same as his, and our, preference for virtue over vice. In each case the former is more pleasant and useful to ourselves and others." Owen, *Hume's Reason*, 222. According to Singer, the Title Principle says that we should allow expedience to determine our beliefs. Singer, "Hume's Extreme Skepticism," 611. No one has emphasized the practical rationality of the Title Principle more than Michael Ridge in his excellent article, "Epistemology Moralized." However, Ridge also says that the Title Principle is a "clearly normative epistemic principle." Ridge, "Epistemology Moralized," 189.

60. W. K. Clifford, "The Ethics of Belief," *Contemporary Review* (January 1877): 346.

61. As Fogelin notes, "Clearly, a philosopher can be a theoretical skeptic of the most general and radical kind without prescribing anything about holding beliefs and without himself following any such prescriptions." *Hume's Skepticism*, 5.

62. Cf. Andrew Chignell, "The Ethics of Belief," *The Stanford Encyclopedia of Philosophy*, Spring 2013 ed., ed. Edward N. Zalta, accessed November 30, 2015, http://plato.stanford.edu/archives/spr2013/entries/ethics-belief, section 2.1

them anyway.[63] Fogelin rightly identifies the Title Principle as a practical rather than epistemological norm.[64] For whatever reason, these commentators do not develop this insight as the key to reconciling Hume's science and skepticism.[65] I aim to do so.

On my reading, Hume champions skepticism in order to show that the epistemic demands of philosophy are not sacred. Life and philosophy sometimes diverge. Philosophy is not the good life and sometimes is not even conducive to the good life. In the final chapter of this book I argue that Hume had reasons to welcome his skeptical conclusions rather than resist them. For one thing, Hume seems to have adopted a moderate approach to his philosophical pursuits after a youthful burnout. The discovery that total commitment to philosophy ends in unlivable suspension of belief would have provided him further confirmation that philosophy deserves only moderate allegiance. Secondly, Hume's skepticism coheres with his irreligion. His idea is that skepticism shows that humans are not creatures of a non-deceiving God, as Descartes thought. Reason is not a divine light or the image of God, and its use does not lead inexorably to flourishing. For Hume, the proper use of our cognitive faculties provides our best chance at navigating safely and perhaps (with luck) even accurately through the world. Certainly, philosophy is safer than superstition. But our cognitive faculties, even used properly, are merely the flawed products of chance in a mindless universe that does not care about us. Skepticism "can have no other tendency than to show the whimsical condition of mankind, who must act and reason and believe; though they

63. Fogelin says that Hume is a theoretical but not a prescriptive skeptic. Fogelin, *Hume's Skepticism*, 6–7. If Hume prescribes a return to the (theoretically unjustified) beliefs of common life and science, presumably he thinks they are practically justified. Singer notes that mitigated skepticism, while it may or may not resolve Hume's epistemological problems, is "a *practical* compromise between extreme skepticism and ordinary belief," and sanctions philosophy just insofar as it is useful and agreeable for some people. Singer, "Hume's Extreme Skepticism," 614–15, 618). Stroud also concludes that although Hume "is in no position to say that profound, careful philosophy is superior" to superstition so far as reliability goes, he "recommends the pursuit of the sceptical or academical philosophy as the best or perhaps the only way to achieve this most natural and therefore most blissful human condition." Stroud, "Hume's Scepticism: Natural Instincts and Philosophical Reflection," in *The Empiricists: Critical Essays on Locke, Berkeley, and Hume*, edited by Margaret Atherton (Lanham, MD: Rowman & Littlefield, 1999), 244, 247. De Pierris discusses the practical value of acknowledging radical skepticism (it wards off theology and metaphysics) and of continuing on with philosophy in spite of radical skepticism. De Pierris, "Hume's Pyrrhonian Skepticism," *Journal of the History of Philosophy* 39, no. 3 (July 2001): 361–63; *Ideas*, 296–306. But recall that she also asserts that ordinary philosophy proceeds at a different level, in terms of different epistemic norms, than radical skepticism. Ibid., 21. None of the above commentators seem to draw much attention to the practical/epistemic divergence or develop it very far.

64. Fogelin, *Hume's Skepticism*, 24

65. In his earlier work, Fogelin emphasizes that Hume's skepticism "clears the ground" (somehow) for his scientific, naturalistic explanations of the mind. Fogelin, *Hume's Skepticism*, 146–51. In his later work, Fogelin emphasizes that Hume's response to radical skepticism—his "mitigated skepticism"—is a psychologically explicable event without any principled rationale. Fogelin, *Hume's Skeptical Crisis*, especially 5–7.

are not able, by their most diligent enquiry, to satisfy themselves concerning the foundation of these operations, or to remove the objections, which may be raised against them" (EHU 12.23). Hume deprives us of a certain kind of personal integrity by denying that our intellectual and practical interests, the demands of philosophy and demands of life, epistemic justification and personal happiness, can fully converge. It is still a provocative message, perhaps most of all to professional philosophers.

I argue that Hume defends the same basic view of epistemology, skepticism, and the proper practical response to skepticism both in the *Treatise* and in the *Enquiry*. The question of continuity between these two works is contentious. Scholars are often reluctant to extend their interpretations of the *Treatise* to cover the *Enquiry* as well. Millican and Qu argue that Hume recognized several philosophical defects in the *Treatise* that made skepticism an insoluble problem; he corrects them in his more mature and polished *Enquiry*.[66] Garrett finds substantive continuity between the two books, but his topical approach does not afford an overall view of each work as an independent whole.[67] I treat the *Treatise* and the *Enquiry* separately for the most part and on their own terms. However, I find that despite differences in detail and mode of presentation, Hume adopts the same basic position in each work. In both works, "philosophy" is his central epistemic term, in both works he brings up skeptical problems to which there is no philosophical solution, and in both works he commends a purely practical response to skeptical dilemmas.

My reading has several advantages. First, it resolves the apparent contradiction between Hume's science and his skepticism. Second, it helps us to sort through Hume's otherwise confusing claims about the skeptical arguments I characterize as *reductios*. Third, it does so in a way that bears out Hume's claim that his philosophical views are largely consistent between the *Treatise* and the *Enquiry*. Fourth, my reading achieves these aims by way of a fresh account of Hume's epistemology which is textually anchored in his idea of "philosophy." Fifth, it offers a historically contextualized explanation of why Hume would want to defend epistemic skepticism.

The rest of this book falls into three main parts. The first part, chapters 1–5, examines the *Treatise*. In chapter 1, I give an overview of Book 1's descriptive account of the mind and its faculties. In chapter 2, I establish Hume's epistemic (philosophical) norms. In chapters 3–4, I show how core beliefs of

66. Millican tracks a number of changes between Book 1 of the *Treatise* and the first *Enquiry* in "Hume's 'Scepticism' About Induction," in *The Continuum Companion to Hume*, ed. Alan Bailey and Dan O'Brien ((New York: Continuum, 2012). He focuses especially on changes in Hume's treatment of skepticism in "Hume's Chief Argument," 95–101. Qu's *Hume's Epistemological Evolution* makes an extended case for the changes in Hume's epistemology, skeptical problems, and response to skepticism between the two works.

67. Garrett, *Hume*.

the senses and reason fall short of these epistemic norms. In chapter 5, I argue that he gives a purely practical response to these skeptical arguments (and to other *reductio* arguments) in *Treatise* 1.4.7. The second part (chapters 6–7) examines the *Enquiry*. Chapter 6 establishes Hume's philosophical norms, which I argue are materially the same as in the *Treatise*. Chapter 7 examines the *Enquiry*'s skeptical arguments (genuine and *reductio*) and argues that Hume again gives a purely practical response to them. In chapter 8 (the third part of this book), I argue that Hume welcomes philosophical skepticism rather than reluctantly accepting it. Skepticism reinforces his moderate approach to philosophy and refutes the doctrine that human cognitive faculties are virtually divine.

On the present interpretation, Hume asks whether the pursuit of philosophy, the rigorous adherence to epistemic norms and the authority of reason, is always the most beneficial course of action. He concludes that it is not. It is in everyone's best interest to hang on to core beliefs, even when they meet with rational defeat. As a matter of psychological fact, we cannot get rid of these core beliefs anyway, even if we try. But the limited practical authority of philosophy does not open up the floodgates of epistemic irresponsibility, superstition, and irrationality. It simply puts philosophy in its proper place—subordinated to human interests and integrated into a well-rounded life.

Chapter One

An Overview of Book 1 of the *Treatise*

This chapter provides an overview of Hume's constructive descriptive claims about the mind in Book 1 of the *Treatise*. It serves several purposes. First, it sketches the broader literary context in which Hume's epistemology and skeptical arguments fit. Succeeding chapters take up specific passages that make less sense without an initial grasp of the overall shape of Book 1. Second, on my reading, Hume bases his epistemology on mental propensities and processes—notably, PIU principles of the imagination. Locating and describing these mental propensities is essential to understanding Hume's epistemic evaluations of them. Third, Hume's skeptical arguments presuppose many of the key findings (for example, about the Copy Principle or external existence) from the constructive part of Book 1 that I introduce here.

This chapter is organized in four parts that correspond to the four parts of Book 1 of the *Treatise*.[1] Neither the four parts of Book 1 nor of this chapter are equal in length. I focus especially on issues that will be relevant to Hume's epistemology and skeptical arguments. As much as possible, I describe the faculties without discussing their normative epistemic status, which I take up in the next chapter.

1. A chapter like this is necessarily, if lamentably, cursory. The footnotes reference only a smattering of the vast secondary literature that takes up these topics in detail. I have given more attention to some of the debate between Garrett and Millican on reason and the understanding because of its direct relevance to topics in succeeding chapters. As will be obvious, I largely line up with Garrett's account of Humean faculties.

OVERVIEW OF PART 1

Impressions and Ideas; Senses, Memory, and Imagination

Hume calls all of the contents of the human mind "perceptions." He sometimes uses the word "perception" to refer to an act of perceiving, but typically uses it to refer to the object of an act of perceiving—that is, to the mental content of an act of perception.[2] Perceptions and their relations comprise the basic subject matter of the entire science of man. The *Treatise* is nothing but an examination of the causes and effects of perceptions (T 1.1.1.6–7).

Perceptions fall into two classes: impressions and ideas (T 1.1.1.1). The superior degree of "force and liveliness" belonging to impressions distinguishes them from ideas. Ideas are "faint images" of impressions. Hume also distinguishes between simple and complex perceptions. Simple perceptions are those that cannot be distinguished into parts, and complex perceptions are those that can (T 1.1.1.2).

Impressions fall into two classes: impressions of sensation and of reflection (T 1.1.2.1). Impressions of reflection arise "in a great measure from our ideas." For example, when we form the idea of a pleasure, we might have a passion of hope or desire to experience that pleasure. Hope and desire are impressions of reflection. Hume defers the investigation of impressions of reflection until Book 2, after he treats the ideas that produce them in Book 1.

The senses are the faculties for receiving impressions of sensation. (Hume refers to "the faculties" of the senses at T 1.4.2.3, 5). An impression of sensation "strikes upon the senses, and makes us perceive heat or cold, thirst or hunger, pleasure or pain of some kind or other" (T 1.1.2.1). He specifies that

> that there are three different kinds of impressions convey'd by the senses. The first are those of the figure, bulk, motion and solidity of bodies. The second those of colours, tastes, smells, sounds, heat and cold. The third are the pains and pleasures, that arise from the application of objects to our bodies. (T 1.4.2.12)

Hume only discusses the senses and sensation insofar as they are accessible by introspection of the mind's own contents. Insofar as the senses and the sensations involve causes, physical or otherwise, outside the mind, he declines to comment upon them, leaving them to the "anatomists and natural philosophers" (T 1.1.2.1).

Hume uses "the senses" with a different meaning in the title of *Treatise* 1.4.2, "Of scepticism with regard to the senses." The skeptical arguments in

2. Ainslie has a sophisticated discussion of this issue. Ainslie, *Hume's True Scepticism*, 53–54, 211.

that section target sensory belief—that is, belief in the existence of objects that continue to exist outside and independently of the mind. But strictly speaking, the senses cannot produce sensory beliefs (T 1.4.2.3–13). The objects of the senses are only impressions insofar as they are currently present to the mind. A different faculty, the imagination, produces belief in the continued and distinct existence of objects. "Of scepticism with regard to the senses" is really about skepticism with regard to the imagination, the faculty that produces sensory beliefs in mind-independent bodies. However, in the section title, Hume uses "the senses" to refer to the faculties that produce belief in body. He sometimes uses "the senses" in this looser way elsewhere as well.[3]

Two distinct faculties produce ideas: the memory and the imagination (T 1.1.3). (Note that Hume sometimes refers to the imagination as "the fancy"). These two faculties produce two "species of ideas." There are two key differences between the ideas that these two faculties produce. The first difference is their relative degrees of force and vivacity. As Hume puts it later, "When I oppose the imagination to the memory, I mean the faculty, by which we form our fainter ideas" (T 1.3.9.19n22). When an impression reappears as an idea of memory, "in its new appearance it retains a considerable degree of its first vivacity" (T 1.1.3.1). By contrast, an idea produced by the imagination "entirely loses that vivacity" of the impression it copies. Three degrees of force and vivacity distinguish impressions (the most forceful and vivid), memorial ideas (the next most forceful and vivid), and imagined ideas (the faintest).

Hume later adds that as the force and vivacity of a memory decay over time, so too it becomes increasingly difficult to determine whether the idea is a memory at all, or something we just imagined (T 1.3.5.5). On the other hand, imagined ideas can sometimes acquire enough force and vivacity that they pass for memories (T 1.3.5.6). "This is noted in the case of liars; who by the frequent repetition of their lies, come at last to believe and remember them, as realities" (T 1.3.5.6).

The second difference between the memory and imagination is that memorial ideas must retain "the same order and form with the original impressions"—unless the memory has "some defect or imperfection" (T 1.1.3.2–3). The imagination, by contrast, has complete liberty "*to transpose and change its ideas*" into any combination that one wishes to contemplate (T 1.1.3.4).

3. For example, he writes that "The only defect of our senses is, that they give us disproportion'd images of things, and represent as minute and uncompounded what is really great and compos'd of a vast number of parts." (T 1.2.1.5). This statement seems to indicate that there is a difference between the impression present to the mind, and the object it represents. But strictly speaking, "our senses offer not their impressions as the images of something *distinct*, or *independent*, and *external* . . . because they convey to us nothing but a single perception, and can never give us the least intimation of any thing beyond" (T 1.4.2.4). So Hume does not appear to be using "our senses" in the strict sense in the sentence quoted from T 1.2.1.5.

Since the imagination can produce ideas in any order and position, it can also produce any complex idea that the memory might produce (T 1.3.5.3). The order and position of ideas does not indicate whether they are products of memory or of imagination.

Although Hume almost always refers to memories as ideas, he sometimes seems to categorize them as impressions, or even to put them in a third category of their own. Here is a typical description of memories as ideas:

> When we remember any past event, the *idea* of it flows in upon the mind in a forcible manner . . . Here then is a sensible difference betwixt one species of *ideas* and another (T 1.1.3.1, italics mine).

More ambiguously, he says that the product of the faculty of memory is "somewhat *intermediate betwixt* an impression and an idea" (T 1.1.3.1, italics mine). Later he says that that "ideas of the memory . . . are *equivalent to* impressions" (T 1.3.4.1, italics mine). In the same paragraph he says that the only thing that can terminate a chain of causal inferences "is an *impression of the memory* or senses" (T 1.3.4.1, italics mine). Similarly, he writes "All our arguments concerning causes and effects" include "an *impression of the memory* or senses" (T 1.3.5.1, italics mine).

I think that the most charitable construal of Hume's meaning is that memories are ideas.[4] They are "somewhat intermediate betwixt an impression and an idea" (T 1.1.3.1) in the sense that their degree of force and vivacity is between that of an impression and that of a merely imagined idea. Memories are "equivalent" to impressions in the sense that they play the same role of establishing the existence of one term of a causal inference (T 1.3.4.1). When Hume speaks of "an impression of memory," he seems to be speaking loosely, describing how the memorial idea faintly copies an impression. When Hume says that causal inference must start from an impression, he means that it must start either from a present impression or from a memorial idea that copies a past impression. Causal inference always derives ultimately from an impression, but it may derive proximately either from a memorial idea or from a present impression.

The Copy Principle

Hume observes that simple impressions and simple ideas exactly resemble one another in content and differ only in their degrees of force and liveliness. "That idea of red, which we form in the dark, and that impression, which

4. This is also Garrett's view. Garrett, *Hume*, 43. According to Millican, by contrast, the memory produces impressions, which are then copied and contemplated as ideas in the imagination. Millican, "Skepticism about Garrett's *Hume*," 221–22n12.

strikes our eyes in sun-shine, differ only in degree, not in nature" (T 1.1.1.5). The same rule does not hold for complex impressions and ideas. On one hand, we have complex impressions for which we have no complex ideas. For example, a person who has had an impression of Paris cannot form an idea of the city that exhaustively corresponds to all the details he originally observed (T 1.1.1.4). Some of the simple impressions that compose the complex impression of Paris have no corresponding simple ideas in the complex idea of Paris. On the other hand, we may have complex ideas for which we have no corresponding complex impression. For example, a person who has never seen the New Jerusalem can nonetheless form the idea of a city with golden streets and ruby walls. However, the simpler component ideas (of a city, of gold, and of ruby) correspond to simpler impressions that the person has had. In short, all ideas are either simple ideas that directly correspond to a simple impression or complex ideas composed of simple ideas that directly correspond to a simple impression.

Hume next concludes to a causal relationship between resembling simple impressions and ideas. He lays it down as a general proposition "*That all our simple ideas in their first appearance are deriv'd from simple impressions, which are correspondent to them, and which they exactly represent*" (T 1.1.1.7). This general proposition has momentous consequences throughout Hume's entire philosophy and is known in the secondary literature as the Copy Principle.[5] Hume argues that the exact correspondence between simple impressions and simple ideas implies some kind of causal dependency of one on the other; the correspondence cannot be mere coincidence. But constant experience shows that simple impressions always precede their simple ideas, whereas simple ideas never precede their corresponding simple impressions. So simple impressions must cause simple ideas (T 1.1.18). Furthermore, those who lack an organ of sensation also lack the corresponding ideas (T 1.1.19). For example, those born blind or deaf lack visual or audial ideas.

Hume admits a possible exception to the Copy Principle. Perhaps someone can form the idea of a shade of a color—say, blue—that they have never seen, if they have seen all the other shades of blue and contemplate the relevant gap in the blue spectrum (T 1.1.1.10). But this is an outlier exception that proves the rule.

Associative Principles and Philosophical Relations

Ideas, Hume observes, do not appear in the imagination in completely random sequences (T 1.1.4). They succeed one another in discernible patterns, according to principles of association. Hume posits three principles of the

5. On the Copy Principle, see especially Garrett, *Cognition* 41–57; Garrett, *Hume* 43–46.

association of ideas in the imagination: resemblance, contiguity (in time or place), and causation. Ideas in the imagination naturally succeed other ideas to which they relate in these three ways. For example, when some person thinks of his mother, it would be natural for him to then think of his mother's parents (his mother's cause), of her other children (her effects), of the aunt that his mother looks like (resemblance), of the last place he saw his mother (contiguity in place), or what else was happening at the time of the visit (contiguity in time). However, these three principles are not absolutely necessary or sufficient for the production of ideas in the imagination (T 1.1.4.1). For example, the man might think of his mother and then of something totally unrelated, such as a unicorn.

Hume transitions from a discussion of the associative principles of the imagination to the complex ideas that arise from the association of simple ideas (T 1.1.4.7). He divides complex ideas into relations, modes, and substances. Relations he treats in *Treatise* 1.1.5, and substances and modes he takes up in *Treatise* 1.1.6.

But before he discusses complex ideas, *Treatise* 1.1.5 introduces another set of relations, which he calls philosophical relations, in distinction from the three natural relations or principles of association introduced previously. He begins by marking the distinction between the two sets of relations:

> THE word RELATION is commonly used in two senses considerably different from each other. Either for that quality, by which two ideas are connected together in the imagination, and the one naturally introduces the other, after the manner above-explained; or for that particular circumstance, in which, even upon the arbitrary union of two ideas in the fancy, we may think proper to compare them. (T 1.1.5.1)

Philosophical relations are the latter kind. Hume identifies seven of them: resemblance, identity, spatial and temporal relations, quantity, quality, contrariety, and causation. For example, I can compare the idea of a red rubber ball with the idea of a mountain and consider whether they resemble one another (resemblance), whether they are identical with one another (identity), their relative age and location (temporal and spatial relations), which is taller (quantity), which is softer (quality), whether they can possibly coexist (contrariety), and whether one caused the other (causation). The idea of a red rubber ball might not stand in any of the three associational relations with the idea of the mountain. That is, the idea of the ball may not naturally introduce the idea of the mountain. But if I voluntarily conceive these two ideas, they might stand in any of these seven philosophical relations.

Modes and Substances

Hume's initial treatment of modes and substances (T 1.1.6) includes a criticism of received accounts of these complex ideas and his own constructive (if somewhat deflationary) account of them. Constructively, Hume says that both these kinds of complex ideas are merely bundles of simple ideas associated by the imagination with each other and with a word that brings them to mind (T 1.1.6.2). What distinguishes substances from modes is a particular kind of false belief that we form about the bundles we call substances (ibid). The simple ideas of the qualities comprising the bundle that we regard as a substance stand in relations of contiguity and causation toward one another. We "suppose" that the qualities are at least inseparable from one another. Commonly we go farther and ascribe their inseparability to their inherence in "an unknown *something*," a substratum.[6] These false suppositions lead us to apply the same term to what is materially a different bundle of simple ideas when we discover new qualities that are also related to the bundle by contiguity and causation (T 1.1.6.3).[7] In the case of modes, we do not form the (false) supposition that these relations *inseparably* connect the constituent qualities. Without this supposition of inseparability, the mind does not assimilate new simple ideas into the complex idea of a mode, without marking the difference by the use of a new name.

On the critical side, Hume uses the Copy Principle to show that we have no idea of an "unknown *something*" in which qualities might inhere, as they do according to the received view of substance. "I wou'd fain ask those philosophers, who found so much of their reasonings on the distinction of substance and accident, and imagine we have clear ideas of each, whether the idea of *substance* be deriv'd from the impressions of sensation or of reflection?" (T 1.1.6.1). The point of this rhetorical question is that we have no impression either of a substratum or of the relation of inherence. But according to the Copy Principle, we therefore have no idea of these items either. So the received doctrine of substance is inconceivable and unintelligible. Later in the *Treatise*, after he introduces the Separability Principle (discussed below), he also argues against the supposition that any two qualities are inseparably related (T 1.4.3, 1.4.5, 1.4.6).

6. Later, in *Treatise* 1.4.3, Hume spells out more of the received doctrine of substance that he is attacking here in 1.1.6. According to that doctrine, a substance is a substratum (which is not a quality) in which accidents (qualities) inhere.

7. Strictly speaking, using the same term to designate two distinct complex ideas is an equivocation. But Hume seems unconcerned about this linguistic convention, so long as we recognize on reflection the truth about our complex ideas. In *Treatise* 1.4.6.21, Hume treats identity ascriptions along these same lines, distinguishing questions of grammatical convention from the philosophical question of identity.

Abstract Ideas; the Separability and Conceivability Principles

Hume's major aim in Section 7, "Of abstract ideas," is to defend George Berkeley's thesis that abstract ideas are particular in the mind's conception of them. Abstract ideas are the ideas whereby we think and reason about a whole class of items. For example, "The abstract idea of a man represents men of all sizes and all qualities" (T 1.1.7.2). The question is whether, when we think and reason about men of all sizes and qualities, we conceive of a man with any particular size or quality. An idea is "general" in the mind's conception of it if the idea represents "no particular degree either of quantity or quality" (T 1.1.7.2). By contrast, an idea is particular in the mind's conception of it if it does represent a particular degree of quantity and quality. Hume holds that "all general ideas are nothing but particular ones, annex'd to a certain term, which gives them a more extensive signification, and makes them recal upon occasion other individuals, which are similar to them" (T 1.1.7.1). We think and reason about (for example) men of all sizes and shapes by conceiving of a man with one particular size and shape. The idea of a man is abstract (or general) insofar as it represents a whole class of items, but particular (not general) in the mind's conception of it—that is, inasmuch as it has fully determinate intrinsic characteristics.

Hume's defense of particular abstract ideas has both critical and constructive components. Critically, after presenting the case in their favor (T 1.1.7.2), he makes three arguments against the existence of ideas that lack particular degrees of quantity or quality (T 1.1.7.3–6). Each argument rests upon a crucial principle that does important work throughout the rest of his corpus. Constructively, he puts forward an original theory of how particular, determinate ideas can represent items with differing characteristics (T 1.1.7.7–10) and responds to two possible objections (T 1.1.7.11–18). Hume says that he places his "chief confidence" in the critical component of his defense, his three arguments for the impossibility of general abstract ideas (T 1.1.7.16). He regards his constructive account of particular abstract ideas as the only available alternative.[8]

Hume's first argument against general abstract ideas (T 1.1.7.3) relies on what is widely known as the Separability Principle.[9] Hume introduces part of the Separability Principle in his discussion of the contrast between memory

8. On Hume's constructive account of abstract ideas, see especially Garrett, *Hume*, 52–60. Millican thinks that Garrett lays more stress on Hume's constructive account than Hume himself does. "Skepticism about Garrett's *Hume*," 207–8. Cf. Garrett's reply, "Millican's 'Abstract,'" 230.

9. See Garrett, *Cognition*, 58–75, and *Hume*, 46–47. According to Millican, the Separability Principle leads to disastrous consequences in the *Treatise*, especially with respect to personal identity, and Hume wisely drops it from the *Enquiry*. "Skepticism about Garrett's *Hume*," 210. Garrett rejoins that although Hume does not explicitly state the Separability Principle in the *Enquiry*, he does not repudiate it. "Millican's 'Abstract,'" 233.

and imagination, where he writes that "Where-ever the imagination perceives a difference among ideas, it can easily produce a separation" (T 1.1.3.4). But his first full statement of the principle comes in the section on abstract ideas:

> We have observ'd, that whatever objects are different are distinguishable, and that whatever objects are distinguishable are separable by the thought and imagination. And we may here add, that these propositions are equally true in the *inverse*, and that whatever objects are separable are also distinguishable, and that whatever objects are distinguishable are also different. For how is it possible we can separate what is not distinguishable, or distinguish what is not different? (T 1.1.7.3)

In sum, the Separability Principle says:

1. Objects are different if and only if they are distinguishable, and
2. Objects are distinguishable if and only if they are separable by the thought and imagination.

Armed with this principle, Hume can argue against non-particular abstract ideas. According to the Separability Principle, if an idea is separable from its quantity and qualities, then it is distinguishable from its quantity and qualities. "But," writes Hume, "'tis evident at first sight, that the precise length of a line is not different nor distinguishable from the line itself; nor the precise degree of any quality from the quality" (T 1.1.7.3). So no idea is separable from its precise degree of quantity and quality; every idea is particular, fully determined in its characteristics.

The second argument against non-particular abstract ideas (T 1.1.7.4–5) invokes the Copy Principle, according to which all ideas are copies of impressions. All impressions, Hume insists, have determinate degrees of quantity and quality. It follows that all ideas have determinate degrees of quantity and quality.

The third negative argument (T 1.1.7.6) relies on what is often called the Conceivability Principle, which Hume introduces here for the first time.[10] According to this initial statement of the Conceivability Principle, "nothing of which we can form a clear and distinct idea is absurd and impossible" (T 1.1.7.6). So if we can form a clear and distinct idea of an object with no precise degree of quantity and quality, then such an object can possibly exist. But, says Hume, it is widely and rightly agreed that indeterminate objects are impossible. It follows that indeterminate ideas are inconceivable.

---

10. Garrett, *Cognition* 24; Garrett, *Hume* 47–49. Garrett thinks that Hume makes conceivability sufficient but not necessary for metaphysical possibility. *Cognition* 257n14; *Hume* 49. On his reading of Hume, there might be non-contradictory ideas that are inconceivable for at least some people, but that represent metaphysical possibilities.

Having shown to his own satisfaction that abstract ideas are particular in the mind's conception of them, he now puts forward an explanation of how particular ideas can represent an infinite range of particular objects (T 1.1.7.7–10). According to Hume, when we discover that objects resemble each other in a certain respect, we apply the same name to all of them, regardless of their other differences. We develop a customary or habitual association between the ideas of these objects and the name they share. The name brings only one of these ideas to mind, but also revives the customary association with the ideas of the other objects bearing that name. The customary association of the general term with the one occurrent idea and with all of the other ideas bearing that general term constitutes the one occurrent idea as an abstract idea. It allows the occurrent idea to represent all the others associated with that name. Whenever our thoughts or assertions about the particular occurrent idea are not true of some of the other ideas associated with the same general term, custom brings these to mind, and we count them as counterexamples. We treat statements about general ideas as true if and only if they are true of all ideas associated with the same general term. For example, if one considers the claim that "All men are short," the particular idea of a man that serves as the abstract idea of men may indeed be of a short man. However, the ideas of tall men, also associated with the general term "man," quickly come to mind and expose the falsity of the claim.

Hume concludes the section by addressing two difficulties. First, he offers more proof for the role of custom by giving four more instances where it operates in a similar way (T 1.1.7.11–16). Second, he explains how, on his theory, it is possible to make a distinction of reason (T 1.1.7.17–18). For example, we can distinguish between the shape and the color of a white globe. A puzzle arises because according to the Separability Principle, if we can distinguish between these two qualities, then they are separable. But shape and color are not separable; we cannot conceive of a shape without color (or tactile qualities), or a colored patch without a shape. Hume thinks he can solve this puzzle with his theory of abstract ideas. On his view, we cannot think of the globe's shape without thinking of its color. However, we can think of its shape and "keep in our eye the resemblance" to other globes (perhaps with different colors) (T 1.1.7.18). We can also think of its color and "keep in our eye" the globe's resemblance to similarly colored objects (perhaps with different shapes). Although we cannot make a real distinction between shape and color, we can distinguish between the various respects in which even a simple object resembles other objects.[11]

11. On distinctions of reason, see Garrett, *Cognition* 62–64; *Hume* 59–60; Donald L. M. Baxter, "Hume, Distinctions of Reason, and Differential Resemblance," *Philosophy and Phenomenological Research* 82.1 (January 2011), 156–82; Taro Okamura, "Hume on Distinctions of Reason: A Resemblance-First Interpretation," *Australasian Journal of Philosophy*, 97.3 (2019), 423–36.

OVERVIEW OF PART 2

In Part 2, Hume deploys the principles of Part 1 to construct a novel account of the ideas of space (or, as he also calls it, extension) and time and to criticize alternative accounts.[12] Hume argues for two main constructive theses in Part 2.[13] First, our ideas of spatial extension and of temporal duration, and the objects of those ideas, are composed of a finite number of simple, indivisible parts (T 1.2.1–2). Second, empty space (i.e., a vacuum) and time without change are inconceivable (T 1.2.3).

After defending these constructive theses, he replies to objections, first to his denial of infinite divisibility (T 1.2.4) and then to his denial of vacuums and changeless duration (T 1.2.5). One of his replies to these objections plays a role in a later discussion of skepticism (EHU 12.18–20). The objection is that certain valid arguments in geometry that start from correct definitions lead to the conclusion that extension is infinitely divisible (T 1.2.4.8). Hume replies that our ideas of geometrical equality and inequality are such that they do not afford reliable inferences about very small, let alone infinitesimal, quantities (T 1.2.4.9–33). Summarizing this point later, he says that the first principles of geometry "are still drawn from the general appearance of the objects; and that appearance can never afford us any security, when we examine the prodigious minuteness of which nature is susceptible" (T 1.3.1.4). Thus the objections to his theory of indivisible parts, drawn from these geometrical demonstrations, have no force.

The final brief section of Part 2 (T 1.2.6) takes up the ideas of existence and of external existence—topics that are tangential to his theory of space and time, but important for his later discussion of external world skepticism. Hume takes it as obvious that whenever we conceive of something, we conceive of it as existing; when we form an idea of an object, we have the idea of its existence (T 1.2.6.1–5). This leads to a disjunctive syllogism. Either the idea of existence is a distinct concomitant of every other idea we have, or else it is identical to every other idea we have. But, he says, there is no idea that is a distinct concomitant of every other idea we have (T 1.2.6.3, 5). Therefore, the idea of existence is identical to every other idea we have.

Next, Hume clarifies what we mean when we talk about an "external existence"—that is, something existing outside of our minds. First, arguing from the Copy Principle again, he eliminates the notion that we might be talking about a non-perception—a "thing specifically different from ideas and

12. For detailed treatments of Part 2, see especially Marina Frasca-Spada, *Space and Self in Hume's* Treatise (New York: Cambridge University Press, 1998), and Donald L. M. Baxter, *Hume's Difficulty: Time and Identity in the* Treatise (New York: Routledge, 2008).

13. Hume provides a helpful outline of his system, and of Part 2, in T 1.2.4.1–2.

impressions" (T 1.2.7.8). Non-perceptions, he argues, are inconceivable: "Now since nothing is ever present to the mind but perceptions, and since all ideas are deriv'd from something antecedently present to the mind; it follows, that 'tis impossible for us so much as to conceive or form an idea of any thing specifically different from ideas and impressions" (T 1.2.6.8). All ideas are copied from other perceptions, and so all ideas represent perceptions.

Hume sounds as if he is softening this conclusion when he writes "The farthest we can go toward a conception of external objects, when suppos'd *specifically* different from our perceptions, is to form a relative idea of them, without pretending to comprehend the related objects" (T 1.2.6.9). However, this "relative" idea is completely empty, a non-idea under a different name. Non-perceptions are simply inconceivable, or as Hume puts it, incomprehensible.

Hume does think that we can conceive of external objects. When we think of them, we think of a certain subset of perceptions. External objects are perceptions that stand in different relations than other perceptions: "Generally speaking we do not suppose them specifically different; but only attribute to them different relations, connections and durations" (T 1.2.6.9). He adds, "But of this more fully hereafter," footnoting "Part 4. Sect. 2." In Part 4, Section 2, he explains that "external objects" are perceptions that continue to exist even when they are not present to a mind—even when they are not among the bundle of perceptions that comprise a mind (T 1.4.2.54). They are unperceived perceptions.

OVERVIEW OF PART 3

Knowledge, Demonstrative Reasoning, and Intuition

Hume begins Part 3 by providing a narrow, technical definition of "knowledge" (or "certainty") (T 1.3.1).[14] Knowledge is the discovery of a relation between ideas by means of intuition or demonstrative reasoning (T 1.3.1.2). Only philosophical relations between ideas, and only four of these, can be objects of knowledge: resemblance, contrariety, degrees in quality, and proportions in quantity or number. These four relations depend exclusively on the intrinsic qualities of the ideas under comparison. The relations are discovered either by intuition or demonstrative reasoning. The first three of these relations can be determined intuitively—that is, immediately upon contemplating the related ideas. However, in some cases, a person cannot immediately

14. As Garrett notes, Hume often uses "knowledge" and "certainty" in a looser, non-technical sense. *Hume*, 42.

determine proportions in quantity or number and must use demonstrative reasoning to do so. Algebra and arithmetic are the sciences in which we can achieve knowledge by way of demonstrative reasoning about quantity or number (T 1.3.1.5). Demonstrative reasoning in geometry does not afford certainty when it comes to very minute quantities, because we have no exact standard of equality in geometry (T 1.3.1.4, 6; cf. T 1.2.4.17–33).

By "demonstrative" (or "abstract") reasoning or argument, Hume seems to mean approximately what contemporary logicians refer to as "deductive" reasoning.[15] According to Hume, the negation of the conclusion of a just, non-fallacious demonstrative argument is inconceivable and impossible (T 13.3.3; 1.3.6.5). But the premises—and therefore the conclusion—of a just demonstrative argument very well may be contingent matters of fact, not simply necessary relations of ideas. So Hume seems to mean that, *assuming the premises are true*, the negation of the conclusion of a just demonstrative argument is inconceivable and impossible. Except for the reference to conceivability, this is the modern definition of deductive validity. Hume also makes a more psychologistic distinction between demonstration and intuition. Intuition grasps its conclusion immediately, at first sight, as it were. Demonstration proceeds in stepwise fashion through a series of premises, intuitively grasping one relation between ideas at a time, until the final conclusion is reached.[16]

Reason and Probable Reasoning

Hume begins his treatment of "probability" with a discussion of reasoning in general.

> ALL kinds of reasoning consist in nothing but a *comparison*, and a discovery of those relations, either constant or inconstant, which two or more objects bear to each other. This comparison we may make, either when both the objects are present to the senses, or when neither of them is present, or when only one. (T 1.3.2.2)

The comparison of objects and discovery of relations between them is a necessary, but not a sufficient, condition for an act of reasoning. In the sentence following the quotation, Hume explains that when both objects are present to the senses, the discovery of the relations is not an act of reasoning but a

15. Millican, "Hume's Sceptical Doubts concerning Induction," in *Reading Hume on Human Understanding*, ed. Peter Millican (New York: Oxford University Press, 2002), 132–36; Qu, *Hume's Epistemological Evolution*, 56n28.

16. Garrett, *Hume*, 92.

passive reception of impressions, better termed "perception."[17] We discover relations between ideas of objects not present to the senses either by an act of intuition or demonstrative reasoning. Like "perception," intuition involves comparison and discovery, but is not an act of reasoning. Intuition discovers relations between ideas "at first sight," "without any enquiry or reasoning" (T 1.3.1.2).[18] Demonstrative reasoning is the discovery of relations between ideas by means of intermediate premises, in contrast to immediate, intuitive discoveries of these relations. The discovery of the relation between an impression of an object present to the senses (or memory of such an impression) and an idea of an object absent from the senses is "probability," elsewhere termed "probable reasoning."[19]

Hume also frequently talks about "reason" in the sense of a faculty, power, or capacity.[20] There is a vigorous debate among commentators about the scope of the faculty of reason. Peter Millican construes reason very broadly as the cognitive faculty, the faculty of truth-apprehension.[21] Reason, on this reading, includes not only demonstrative and probable reasoning, but also intuition, the senses, and memory.[22] I am persuaded however by Don Garrett's arguments that Hume uses "reason" in a much narrower sense to refer only to the capacity for inference—that is, for probable and demonstrative

17. Note that here Hume uses the word "perception" to refer to a mental event rather than to the object of a mental event.

18. Although Hume excludes intuition from reason in this paragraph, Millican argues that typically, Hume implicitly includes intuition when he talks about reason, and that intuition belongs to the faculty of reason. Millican, "Skepticism about Garrett's *Hume*," 211, 215. This claim is part of Millican's case against Garrett's position that reason is only the faculty of inference. Garrett agrees that Hume often implicitly refers to intuition along with reason. However, Garrett thinks intuition does not belong to reason, although it does belong to the understanding. Garrett, "Millican's 'Abstract,'" 235–36. I can agree with both that Hume often also has intuition in mind too when he talks about reason. I incline toward Garrett's classification scheme, but I do not think that much rides on it for my purposes. Even if Millican is right that intuition belongs to reason, it does not follow that reason is as broad as Millican takes it to be. I discuss this debate about the scope of reason below.

19. This classification of demonstration and probability is implicit in *Treatise* 1.3.2.2. Hume only uses the word "probability" in the title of *Treatise* 1.3.2. He does not use the word again until *Treatise* 1.3.6, where it is interchangeable with "probable reasoning" (T 1.3.6.6).

20. Commentators typically take reason to be a faculty. See for example Garrett, *Hume*, 88; Millican, "Hume's 'Scepticism,'" 79–83. However, although Hume refers to a "reasoning faculty" (T Intro. 5; 1.3.15.12; ATHT 3), he does not directly mention a "faculty of reason" in Book 1 of the *Treatise*. The character Philo does mention the "faculty of reason" (D 1.3).

21. Ibid., 82–83. He notes that Hume's use of "reason" might nuance this meaning in different contexts. "Reason" might refer to truth-apprehension, "*however well and by whatever processes it operates*," or to "*processes that are commonly taken to be involved in truth-apprehension*," or to "processes that *operate successfully to apprehend truth*," or to "*the faculty of truth-apprehension acting entirely alone*," or to the product of this faculty. Ibid.

22. Millican seems to include the senses (along with memory and intuition) as specific sub-faculties of the general cognitive faculty known as reason. "Hume's 'Scepticism,'" 83. Elsewhere he says that the senses "report to" reason. "Skepticism about Garrett's *Hume*," 223n24. But as Garrett points out, "reporting to" is not the same as "belonging to," and in fact implies a distinction between reason and the senses. "Millican's 'Abstract,'" 237–38.

reasoning.[23] Garrett gives three textual reasons for favoring a narrower view of "reason." First, in an important footnote, Hume identifies "reason" with probable and demonstrative reasoning:

> When I oppose the imagination to the memory, I mean the faculty, by which we form our fainter ideas. When I oppose it to reason, I mean the same faculty, excluding only our demonstrative and probable reasonings. (T 1.3.9.19n22)[24]

Second, *Treatise* 1.3.16, "Of the reason of animals," is specifically about the inferential "reasoning faculty" of animals (T 1.3.15.12). Third, at three crucial junctures (T 1.3.6, 2.3.3, and 3.1.1), Hume argues that if demonstrative and probable reasoning cannot produce something, then reason alone cannot produce it.[25] If Hume construes reason to include also the senses and memory, then he would need to argue that they cannot produce the items in question either. On the contrary, Hume consistently contrasts reason with the senses.[26]

The objects of probable reasoning are the other three philosophical relations not discovered by demonstration: identity, temporal and spatial relations, and causation (T 1.3.2.1, 3). However, probable reasoning only directly discovers causal relations. We can discover spatiotemporal or identity relations by direct simultaneous observation of both relata, but in those cases, the discovery is not reasoning at all, but perception. If one relatum is not directly present to the senses, then we can only discover spatiotemporal or identity relations by means of discovering a causal relation between the relata. So the only relation that is a direct object of probable reasoning is the causal relation. Spatiotemporal and identity relations between impressions and ideas are indirect objects of probable reasoning.

The Understanding

At this point I digress from following the order of topics as they appear in the *Treatise* to give an initial sketch of the understanding.[27] The understanding is the titular subject of all of Book 1. On this basis, it is plausible to think that

---

23. *Hume*, 88–92; "Millican's 'Abstract,'" 236–38. Garrett admits that immediate intuition is a challenge for his classificatory scheme.

24. Millican admits that the footnote seems to restrict reason to probable and demonstrative inference, but says that it must be "strictly inaccurate," since it would exclude immediate, non-inferential intuition from reason too. "Skepticism about Garrett's *Hume*," 211. But Hume does explicitly say that immediate intuition occurs without reasoning (T 1.3.1.2), so it does not seem to be a lapse.

25. *Hume* 92; "Millican's 'Abstract,'" 237.

26. Garrett, "Millican's 'Abstract,'" 237–38. Garrett cites as examples the difference between Hume's skepticism with regard to reason (T 1.4.1) and skepticism with regard to the senses (T 1.4.2), his question of whether an opinion is produced by "the *senses, reason*, or the *imagination*" (T 1.4.2.2), and his allusion to qualities of which we are informed by "neither sense nor reason" (EHU 4.16).

27. Hume does not devote any distinct section to the understanding. Since it takes shape largely in relation to other faculties, it is easier to discuss it once they are already on the table.

the understanding deals with ideas (the subject of Parts 1–2) and the discovery of their relations through knowledge and probability (the subject of Part 3). Hume's first mention of the understanding in the body of the text seems to confirm this presumption: "'Tis impossible to tell what changes and improvements we might make in these sciences were we thoroughly acquainted with the extent and force of human understanding, and cou'd explain the nature of the ideas we employ, and of the operations we perform in our reasonings" (T Intro. 4). The human understanding is directly relevant to "the nature of ideas" (the topics of Parts 1–2) and "the operations we perform in our reasonings" (the topic of Part 3).

In a long footnote, Hume says that conception, judgment, and reasoning are all acts of the understanding (T 1.3.7.5n20). In a sense, he says, these three acts are all just variations of one act, conception. Conception is a necessary but not sufficient element of reasoning and judgment; we can conceive of an object without reasoning or forming a judgment about its reality. Reasoning is necessary but not sufficient for judgment; reasoning might lead to a suspension of judgment. Conception, judgment, and reasoning correspond to the topics of the main constructive parts of Book 1: Parts 1–2 deal with conception (ideation in general), and Part 3 deals with reasoning and judgment.

These observations about the understanding provide clues to its relationship to other faculties. First, the understanding at least overlaps with the imagination, insofar as Hume describes them both as the faculty of ideas and conception.[28] Furthermore, the understanding is no wider than the imagination; Hume denies that we have any ideas that "fall not under the conception of the fancy" (T 1.3.1.7). Second, the memory is also a faculty of ideas, so the understanding may include it as well as the imagination.[29] However, Hume does not seem to ascribe "conception" to the memory as he does to the imagination. As noted above, he treats memorial ideas as somewhat of a special case. So it is less clear whether Hume classifies it as belonging to the understanding.

Third, the understanding includes reason but is broader than reason. Millican argues that Hume uses "reason" and "understanding" interchangeably because they name one and the same faculty—the cognitive faculty, broadly construed.[30] Garrett too acknowledges that Hume does often use "understand-

---

28. Here are a few examples of Hume ascribing conception to the imagination. The mathematicians are wrong to claim that mathematical ideas "fall not under the conception of the fancy" (T 1.3.1.7). Hume uses the word "fancy" interchangeably with "imagination" (cf. T 1.1.3.4; 1.1.4.2). So in this passage the fancy, the imagination, is the faculty of conception. Other passages confirm that the imagination is the faculty of conception: "The imagination . . . interposes not a moment's delay betwixt the hearing of the one, and the conception of the other" (T 1.3.6.14); "the imagination is free to conceive" (T 1.3.7.3).

29. Garrett thinks that the memory belongs to the understanding. Garrett, *Hume* 348.

30. Millican, "Hume's 'Scepticism,'" 80; "Skepticism About Garrett's *Hume*," 214–16.

ing" interchangeably with "reason," or to refer to reason together with intuition.[31] However, he thinks that Hume also uses "understanding" in a broader sense to refer to the faculty of perceiving or having ideas.[32] There is a good explanation, I think, for why Hume sometimes seems to use "understanding" and "reason" virtually interchangeably, though the former is broader than the latter. Hume might very well use "understanding" and "reason" interchangeably in contexts where reason is the only means whereby the understanding could perform some act—for example, the formation of a belief about an unobserved matter of fact. So far as that kind of belief-formation goes, understanding and reason are coextensive.

Fourth, the understanding and the senses are mutually exclusive faculties. The understanding is a faculty of conception—that is, of ideas. Strictly speaking, the senses are faculties for receiving impressions. Even if we construe "the senses" simply as "whatever processes produce belief in continued and distinct existences," as Hume uses the term in the title of *Treatise* 1.4.2, the senses still do not belong to the understanding. The principles that produce belief in body do not belong to reason, and Hume concludes from that fact that they do not belong to the understanding: "This sentiment, then, as it is entirely unreasonable, must proceed from some other faculty than the understanding" (T 1.4.2.14).

The Outline of Part 3

After giving an initial sketch of the domains of knowledge and probability, Hume turns his attention to the nature of the causal relation and its relationship to probable reasoning. In order to clarify the idea of causation, he looks for its source impression: "LET us therefore cast our eye on any two objects, which we call cause and effect, and turn them on all sides, in order to find that impression, which produces an idea of such prodigious consequence" (T 1.3.2.5). It is not the impression of any quality that produces the idea of causation (T 1.3.2.5). Every existent thing is either a cause or effect, says Hume, but no quality is shared by all existents. He concludes that the idea of causation derives from an impression of a relation, not of a quality (T 1.3.2.6). He identifies three relations that seem to hold between causes and effects: spatial contiguity, temporal contiguity, and temporal succession (causes immediately precede effects) (T 1.3.2.6).[33] But even if these three relations are necessary conditions of causation, they are not sufficient conditions. "An object may be

31. *Hume*, 89, 348.

32. Garrett, *Hume* 88–91, 348.

33. Hume makes these observations tentatively, however, and says "the affair is of no great importance" (T 1.3.2.8). As he forecasts in footnote 16, he qualifies the claim about the spatial contiguity of causes in T 1.4.5.

contiguous and prior to another, without being consider'd as its cause. There is a NECESSARY CONNEXION to be taken into consideration; and that relation is of much greater importance, than any of the other two above mention'd" (T 1.3.2.11). But Hume can find no other impression that could be the source of the idea of the necessary connexion between cause and effect (T 1.3.2.9, 12).

He proposes to investigate two adjacent questions about causation in the hopes that an answer to the main question about the necessary connection will emerge (T 1.3.2.13–15). The rest of Part 3 is structured around these adjacent questions and the light they shed on the idea of causation. The first question is "For what reason we pronounce it *necessary*, that every thing whose existence has a beginning, shou'd also have a cause?" (T 1.3.2.14), and Hume addresses it in *Treatise* 1.3.3. The second question, "Why we conclude, that such particular causes must *necessarily* have such particular effects," divides into two sub-questions: (a) "what is the nature of that *inference* we draw from the one [causal relatum] to the other," and (b) what is the nature "of the *belief* we repose in it [i.e., in the other, inferred causal relatum]" (T 1.3.2.15). Hume addresses (a) in *Treatise* 1.3.4–6, and (b) in *Treatise* 1.3.7–13. He finally answers the question about the idea of necessary connection in *Treatise* 1.3.14.

The Causal Maxim

According to Hume, most philosophers take it as intuitively certain that "*whatever begins to exist, must have a cause of existence*" (T 1.3.3.1). In the next paragraph, he expands this maxim to include not only "every new existence" but also every change, that is, "new modification of existence" (T 1.3.3.2). For convenience, I refer to the principle "Whatever begins to exist, and every change, must have a cause" as the Causal Maxim.

Hume denies that the Causal Maxim is an object of knowledge (in the strict sense) or certainty, whether intuitive or demonstrative. In the first place, the Causal Maxim is not one of the four relations that can be known by intuition. But secondly, Hume has a general argument to show that we have no knowledge of the Causal Maxim. If something is an object of knowledge (that is, an object of intuition or demonstration), then its negation is impossible. But the negation of the Causal Maxim is not impossible. To show that the negation of the Causal Maxim possible, Hume invokes the Separability and Conceivability Principles. We can distinguish between a cause and an effect, so therefore we can also conceive of one without the other (by the Separability Principle). But if we can conceive of an effect (a change, or an existence-beginning) without a cause, then it is possible (by the Conceivability Principle). Therefore, the Causal Maxim is possibly false and hence not an object of intuition or demonstration, nor of knowledge in the strict sense. Hume goes on to show what he takes to be the specific fallacies in demonstrative arguments offered in support of the Causal Maxim (T 1.3.3.4–8).

Hume does not deny that Causal Maxim is true or even that we can have a philosophically justified belief in it, but he does deny that it is an object of knowledge in his strict sense—that is, a relation of ideas known by intuition or demonstrative reasoning. If belief in the Causal Maxim does not arise from intuition or demonstration, then it must arise from experience and observation. But instead of investigating the experiential origins of belief in the Causal Maxim in particular, he instead turns to the question of the experiential origins of any belief about causal relations, and of why we infer any one causal relatum from another (T 1.3.3.9). The explanation of causal inference is also the explanation of the belief in causal relations, and thus of belief in the Causal Maxim.

Causal Inference and Unreasoning Imagination

Hume takes up the first part of his second question, about the nature of causal inference (T 1.3.2.15), in *Treatise* 1.3.4–6. By "causal inference" I mean an inference from one observed object to another unobserved object, based on the presumption that the two objects are causally related. Hume does not use the adjective "causal." However, I am using "causal inference" for ease of utterance to refer to what he variously calls "the inference from the impression to the idea" (section title of T 1.3.6), "the inference we draw from cause to effect" (T 1.3.6.1), "the transition from an impression present to the memory or senses to the idea of an object, which we call cause or effect" (T 1.3.6.4), "probable reasonings" (T 1.3.6.6), and similar locutions. These expressions refer to what is now called inductive inference—a term I will sometimes use, although Hume does not. I distinguish between causal inference and another mental operation whereby we come to believe or presume that two objects are causally related. For Hume, the presumption that a causal relation exists between two objects precedes and makes possible the causal inference from one object to another. It is only after "we call the one *cause* and the other *effect*" that we "infer the existence of the one from that of the other" (T 1.3.6.2).[34]

Hume begins by making some preliminary observations about causal inferences (T 1.3.4–5). All causal reasoning that issues in belief in a real existent must begin from an immediate impression or memory (T 1.3.4.1). It is possible to make a causal inference based on the conclusion of another causal

34. Garrett draws a similar distinction between "causal reasoning" and "causal judgment." Don Garrett, "Hume's Theory of Causation: Inference, Judgment, and the Causal Sense," in *The Cambridge Companion to Hume's* Treatise, edited by Donald C. Ainslie and Annemarie Butler (New York: Cambridge University Press, 2015), 69–100.

inference, but eventually the inference chain must terminate in a present impression or memorial idea. Any inference we make from a mere unbelieved idea is hypothetical; we do not believe the conclusion unless we believe in the premise from which it is inferred (T 1.3.4.2). The present impressions and memories from which we make causal inferences are beliefs; their superior force and vivacity distinguishes them from mere ideas of the imagination (T 1.3.5.7).

Hume now makes what is perhaps his most famous argument for his most famous conclusion: that causal inference is not produced by the understanding but by the imagination (T 1.3.6). Here is how he frames the question:

> Since it appears, that the transition from an impression present to the memory or senses to the idea of an object, which we call cause or effect, is founded on past *experience*, and on our remembrance of their *constant conjunction*, the next question is, Whether experience produces the idea by means of the understanding or imagination; whether we are determin'd by reason to make the transition, or by a certain association and relation of perceptions. (T 1.3.6.4)

He goes on to argue that experience does not produce causal inference by means of the understanding or reason, but by means of imagination and a certain association of perceptions. I unpack the details of this argument in the next chapter. The question I wish to take up now is terminological: what is the relationship between understanding, reason, imagination, and "association and relation of perceptions"?

Although in one sense the understanding belongs to the imagination, the passage quoted above (T 1.3.6.4) shows that in another sense these faculties are mutually exclusive. If the idea of a causal relatum is produced by the understanding, then it is not produced by the imagination, and vice versa. Reason and the associative principles of the imagination are mutually exclusive. But this is puzzling, since the understanding, and reason with it, belongs to the imagination.

The explanation of the puzzle is that Hume is using "imagination" in a special, restricted sense. He explicitly articulates the relevant distinction between broad and narrow senses of "imagination" in another important footnote:

> In general we may observe, that as our assent to all probable reasonings is founded on the vivacity of ideas, it resembles many of those whimsies and prejudices, which are rejected under the opprobrious character of being the offspring of the imagination. By this expression it appears that the word, imagination, is commonly us'd in two different senses; and tho' nothing be more contrary to true philosophy, than this inaccuracy, yet in the following reasonings I have often been oblig'd to fall into it. When I oppose the imagination to the memory, I mean the faculty, by which we form our fainter ideas. When I op-

> pose it to reason, I mean the same faculty, excluding only our demonstrative and probable reasonings. When I oppose it to neither, 'tis indifferent whether it be taken in the larger or more limited sense, or at least the context will sufficiently explain the meaning. (T 1.3.9.19n22)

In one sense, the imagination is the faculty for fainter ideas. It encompasses the understanding, includes reasoning (demonstrative and probable), and excludes memory. Following Garrett, I will use the term "inclusive imagination" with this sense, since imagination in this sense includes reason.[35] In another sense, imagination refers to the same faculty, but exclusive of demonstrative and probable reasonings. Again, following Garrett, I will use "unreasoning imagination" with this sense.

This distinction between inclusive and unreasoning imagination illuminates the question Hume poses in *Treatise* 1.3.6.4. His question there is whether the transition from a present impression to the idea of a causal relatum is produced by reasoning or by the principles of unreasoning imagination. He answers that causal inference is produced by principles of the unreasoning imagination, namely the principle of custom (T 1.3.7.6; 1.3.8.10). Causal inference is an act of reasoning, that is, an act of understanding, which excludes unreasoning imagination. However, that act of reasoning is caused by custom, a principle that belongs to the unreasoning imagination, which excludes the understanding.

Hume later draws another distinction between two classes of principles of the imagination, one of which is "received by philosophy" and the other which is "rejected" by philosophy:

> In order to justify myself, I must distinguish in the imagination betwixt the principles which are permanent, irresistable, and universal; such as the customary transition from causes to effects, and from effects to causes: And the principles, which are changeable, weak, and irregular; such as those I have just now taken notice of. The former are the foundation of all our thoughts and actions, so that upon their removal human nature must immediately perish and go to ruin. The latter are neither unavoidable to mankind, nor necessary, or so much as useful in the conduct of life; but on the contrary are observ'd only to take place in weak minds, and being opposite to the other principles of custom and reasoning, may easily be subverted by a due contrast and opposition. For this reason the former are received by philosophy, and the latter rejected. (T 1.4.4.1)

Philosophy approves of "principles of the imagination, which are permanent, irresistable, and universal." (Henceforth, I will refer to these as PIU principles). Philosophy disapproves of those principles that are "changeable, weak,

35. Garrett, *Hume*, 88.

and irregular." (Henceforth, I will refer to these as CWI principles.) These contrasting sets of principles play a central role in my account of Hume's theory of justification, and I discuss their normative significance at length in chapter 2. For now, I only want to give an initial description of what these principles do and do not include.

Peter Millican argues that all and only PIU principles of the imagination are "appropriately dignified with the name of *reason* or the *understanding*."[36] This follows from his broader view, discussed above, that "reason" does not refer simply to one of our cognitive faculties, but our overall cognitive faculty.[37] Millican also argues that the distinction Hume draws in *Treatise* 1.3.9.19n22 (between inclusive and unreasoning imagination) is the same as the distinction he draws in *Treatise* 1.4.4.1 between PIU and CWI principles of the imagination.[38] On this view, unreasoning imagination is opposed to all PIU principles.

But the PIU/CWI distinction (T 1.4.4.1) does not, I argue, coincide with the distinction between the inclusive and unreasoning imagination (T 1.3.9.19n22). First, probable reasoning is an act of the understanding that is caused by custom, which is a principle of the unreasoning imagination. But the "principles of custom and reasoning" are paradigmatic examples of PIU principles. Reason belongs to the understanding and custom does not, but both are PIU principles of the imagination. Second, Hume does not say that custom and reason are the only PIU principles of the imagination. He simply lists custom and reason as paradigmatic examples of PIU principles ("such as").[39] Third, Millican holds that belief in body is produced by PIU principles, a position for which I also argue in chapter 4.[40] But Hume explicitly says that reason does not produce belief in body (T 1.4.2.14). So PIU principles are broader than reason, extending also to those principles of the unreasoning imagination that produce belief in body.[41]

Millican reads Hume as identifying the understanding with PIU principles in a sentence from "Conclusion of this book" (T 1.4.7).[42] The sentence reads:

> But on the other hand, if the consideration of these instances makes us take a resolution to reject all the trivial suggestions of the fancy, and adhere to the understanding, that is, to the general and more establish'd properties of the

36. Millican, "Hume's 'Scepticism,'" 84.
37. Ibid. 83–85.
38. Ibid. 84, 102n121.
39. In chapter 4, I argue specifically that according to Hume, the principles that produce the *vulgar* belief in body are PIU principles.
40. Millican, "Skepticism about Garrett's *Hume*," 222n17.
41. Cf. Garrett, "Millican's 'Abstract,'" 233–35.
42. Millican, "Hume's 'Scepticism' About Induction," 84–85; "Skepticism about Garrett's *Hume*," 212.

> imagination; even this resolution, if steadily executed, wou'd be dangerous, and attended with the most fatal consequences. (T 1.4.7.7)

The "that is" in this passage seems to strictly identify the understanding with "the general and more establish'd properties of the imagination." It is initially plausible to think, as Millican does, that these "general and more establish'd properties" are the PIU principles.

However, there are good reasons to resist this reading of the sentence. First, Millican's reading faces the problems noted above, that custom and the processes producing sensory beliefs are PIU principles that do not belong to reason or the understanding. Furthermore, it is not necessary to read the sentence as stating that *only* the understanding contains PIU principles. Plausibly, Hume means that the understanding is *among* the PIU principles. In context, he has in view the skeptical argument from the self-subversion of reason (T 1.4.1). His point is that the deliverances of the understanding (more specifically, of reason) have a claim to total adherence because they are PIU principles, and as such carry all the normative authority of philosophy (as I argue in chapter 2). In other words, his point is that the understanding belongs to the PIU principles, not that the PIU principles all belong to the understanding.

With these distinctions in hand, Hume's conclusion in *Treatise* 1.3.6 about the causes of causal inference comes into clearer focus. He asks whether experience produces causal inference by means of reasoning (that is, by means of an act of the understanding) or by non-rational associative principles that belong to unreasoning imagination. He answers that experience produces causal inferences by means of unreasoning imagination. However, the principles of unreasoning imagination that produce causal inference are PIU principles, as he will later say.

This account raises another puzzle about the relationship of causal inference to the understanding. Causal inference is the same as probable reasoning, and Hume says that all reasoning is an act of the understanding (T 1.3.7.5n20). But in *Treatise* 1.3.6 he claims that causal inference (an act of reasoning) is not caused by the understanding, but by a principle of the unreasoning imagination. This seems to be a contradiction.

However, the contradiction is merely apparent. Causal inference (probable reasoning) is an act of the understanding that is caused by a principle outside the understanding.[43] In fact, several faculties play a role in producing the act of causal inference. The senses supply present sense impressions, memory supplies the recollection of similar past instances, and unreasoning imagination supplies the associative principle of custom. All of these faculties contribute to what is finally an act of the understanding—causal inference.

43. Garrett, *Hume*, 178–79.

The Nature of Belief

Hume turns from his examination of causal inferences to an examination of the beliefs that such inferences produce as their conclusions. He argues that what distinguishes the "simple conception" of an object from a belief in its existence is that a belief is a more forceful and vivid idea than a simple conception (T 1.3.7).[44] A simple conception and a belief have the same content; their only difference "must lie in the *manner*, in which we conceive it [the object]" (T 1.3.7.2). This leads to Hume's official definition of a belief as "A LIVELY IDEA RELATED TO OR ASSOCIATED WITH A PRESENT IMPRESSION" (T 1.3.7.5).

Somewhat confusingly, this "official definition" of belief only fully applies to one kind of belief. It is tailor-made to capture beliefs produced by causal inference, which are lively ideas associated by custom with a present impression (or memory). But Hume recognizes many beliefs that are not always related to or associated with a present impression, such as memories, the perception of relations between present impressions, and beliefs about relations of ideas resulting from intuition or demonstrative reasoning. His official definition of belief is not intended to exclude these others.

In other contexts, Hume makes a high degree of force and vivacity sufficient to constitute an idea as a belief, dropping the required relation to a present impression. For example, in a footnote appended to the paragraph containing the official definition, he describes a belief as "only a strong and steady conception of any idea, and such as approaches in some measure to an immediate impression" (T 1.3.7.5n20). Beliefs produced by memory and immediate sensation fall under this broader definition. Hume seems to be using this broader definition when he writes that "the *belief* or *assent*, which always attends the memory and senses, is nothing but the vivacity of those perceptions they present; and that this alone distinguishes them from the imagination" (T 1.3.5.7). It is less clear whether force and vivacity are even necessary conditions for belief in relations of ideas. In a passage where he directly addresses this category of belief, he omits any mention of the intrinsic feeling, force, or vivacity of the ideas and instead emphasizes that we cannot conceive of the ideas in any other relation (T 1.3.7.3).[45]

44. In the Appendix, Hume adds a paragraph amending this account of belief to say that what distinguishes it from simple conception is not force and vivacity, but a feeling in the mind (T 1.3.7.7App). No single word quite captures this feeling, says Hume, but it is familiar to everyone from their own experience. The important point for Hume is that the difference between simple conception and belief is not a matter of the content of ideas but the manner of conception. The term we use to describe that distinct manner is less important.

45. "Wherein consists the difference betwixt believing and disbelieving any proposition? The answer is easy with regard to propositions, that are prov'd by intuition or demonstration. In that case, the person, who assents, not only conceives the ideas according to the proposition, but is necessarily determin'd to conceive them in that particular manner, either immediately or by the interposition of other ideas. Whatever is absurd is unintelligible; nor is it possible for the imagination to conceive any thing contrary to a demonstration." (T 1.3.7.3)

Causes of Belief

Custom (or habit) is the principle of the imagination that calls up the idea of a wonted conjunct and gives it the force and vivacity sufficient for belief (T 1.3.8.7–17). Hume gives an official definition of custom: "we call every thing CUSTOM, which proceeds from a past repetition, without any new reasoning or conclusion" (T 1.3.8.10). Because of custom or habituation, when we have had repeated experience of two relata, we instinctively believe in one upon observing the other. All causal reasoning depends on this principle of custom.

But custom can also produce beliefs in a different way than causal reasoning, in a process that Hume calls "EDUCATION" (T 1.3.9.16–19). What he means is closer to what contemporary people might call indoctrination or brainwashing. It involves the inculcation of belief by means of sheer repetition of a single idea. In "education,"

> a mere idea alone, without any of this curious and almost artificial preparation [i.e., past experience of constant conjunction, and the present impression of one conjunct], shou'd frequently make its appearance in the mind, this idea must by degrees acquire a facility and force; and both by its firm hold and easy introduction distinguish itself from any new and unusual idea. (T 1.3.9.14)

Both operations of custom cause the mind to transition easily to an idea and lend that idea more force and vivacity than others. In education, sheer repetition of an idea renders it sufficiently familiar and forceful to count as a belief. When we have heard a claim made often enough—even when we have stated a lie often enough—we start to believe it, sometimes very strongly.

Hume finds confirmation for his theory of causal inference and belief formation in analogous mental phenomena. Just as one kind of custom can produce belief in a causal relatum, so also a second kind of custom, "education," can produce belief in an oft-repeated idea. Just as a causal relation between an impression and an idea can vivify the idea (T 1.3.8.6), so too can the other two associative relations of resemblance (T 1.3.8.3–4) and contiguity (T 1.3.8.5). Hume subsumes these phenomena under "a general maxim in the science of human nature, *that when any impression becomes present to us, it not only transports the mind to such ideas as are related to it, but likewise communicates to them a share of its force and vivacity*" (T 1.3.8.2). However, causation is the only associative principle that can vivify an idea enough to make it a belief (T 1.3.9.1–15). The other two associative principles of resemblance and contiguity, operating by themselves, do not lend ideas enough force and vivacity to constitute them as full-fledged beliefs. Hume also finds confirmation for his theory of belief in the fact that beliefs and impressions stand in similar causal relationships to the passions (T 1.3.10).

The Judgment

The faculty of judgment is a sub-faculty of the understanding that produces the beliefs resulting from causal inference (T 1.3.9.3). The judgment plays a leading role in Hume's discussion of belief in the latter half of Part 3.[46] Hume can use the word "judgment" to refer either to a faculty of belief-formation or to a belief itself.[47] Other faculties besides the judgment can also produce judgments and beliefs.[48] But only the judgment produces beliefs from probable reasoning.

Hume also sometimes uses "judgment" in a wider sense. For example, he writes that the process of "education" influences the judgment to form beliefs (T 1.3.9.16–19). Education is not a process of probable reasoning. However, in this case, Hume prefers to ascribe the belief to the imagination rather than the judgment. He writes that

> As liars, by the frequent repetition of their lies, come at last to remember them; so *the judgment, or rather the imagination*, by the like means, may have ideas so strongly imprinted on it, and conceive them in so full a light, that they may operate upon the mind in the same manner with those, which the senses, memory or reason present to us. (T 1.3.9.19, italics mine)

The ideas imprinted on "the judgment, or rather the imagination" by education are distinct from those presented by senses, memory, or reason, but they are equally vivid and forceful. These ideas of education belong to the judgment only in a broad sense. Hume prefers not to ascribe them to the judgment, but to the imagination, as he signals by "or rather the imagination." I use the term "judgment" in Hume's preferred way to refer only to the faculty that produces beliefs from probable reasonings, unless otherwise specified.

The contrast between judgment (in Hume's preferred narrower sense) and imagination correlates with the contrast between reason and unreasoning imagination. In the footnote appended to the paragraph quoted above, Hume distinguishes between inclusive and unreasoning imagination (T 1.3.9.19n22). Since judgment requires probable reasoning, judgment and

---

46. The claim that the judgement is a sub-faculty of the understanding follows from the fact that judging, like conception and reasoning, is an act of the understanding (T 1.3.7.5n20). Millican, by contrast, thinks that "the judgment"—like reason—is equivalent to "the understanding." "Skepticism about Garrett's *Hume*," 222n13.

47. Hume refers explicitly to the judgment as a faculty at various points (for example, "that faculty," T 1.3.9.16; "the former [faculty]," 1.3.13.11; "that latter faculty," T 1.3.13.13). He uses "judgment" as a synonym for "opinion" and "belief" when he writes, for example, that "Thus it appears upon the whole, that every kind of opinion or judgment, which amounts not to knowledge, is deriv'd entirely from the force and vivacity of the perception, and that these qualities constitute in the mind, what we call the BELIEF of the existence of any object" (T 1.3.13.19).

48. For example, Hume writes about "judgments" of the senses and imagination (T 1.2.4.23–24; 1.3.1.4, 6).

unreasoning imagination are mutually exclusive. The maxims of education, as well as many other "whimsies and prejudices," belong to unreasoning imagination, not to the judgment.[49]

Hume lays down eight general "Rules by which to judge of causes and effects" that constitute norms for the faculty of the judgment (T 1.3.15). "We ought to regulate our judgment concerning causes and effects" by these rules of logic (T 1.3.13.11). They are easy to list, but their application "requires the utmost stretch of human judgment" (T 1.3.15.11). Hume ascribes inferences that follow these general rules to the faculty of judgment; prejudicial inferences that violate these rules he ascribes to the imagination (T 1.3.13.11).

Treatise 1.3.11–13 all discuss how different varieties of probable reasoning influence the faculty of judgment. Although Hume has so far provisionally divided human reason into only two categories, knowledge and probability, he now proposes a third category, proof (T 1.3.11.2). Proof includes arguments from causation that exceed mere probability. He defines the three categories as follows:

> By knowledge, I mean the assurance arising from the comparison of ideas. By proofs, those arguments, which are deriv'd from the relation of cause and effect, and which are entirely free from doubt and uncertainty. By probability, that evidence, which is still attended with uncertainty. (T 1.3.11.2)

Treatise 1.3.11–13 take up probability in this narrower sense. These sections provide psychological accounts of several different kinds of probable judgments, the details of which are not essential for present purposes.

The Idea of Necessary Connection

Hume finally identifies the impression from which the idea of necessity is copied toward the end of Part 3 (T 1.3.14). After frequently observing a conjunction of two objects that are contiguous and successive,

> I find, that upon the appearance of one of the objects, the mind is *determin'd* by custom to consider its usual attendant, and to consider it in a stronger light upon account of its relation to the first object. 'Tis this impression, then, or *determination*, which affords me the idea of necessity. (T 1.3.14.1)

Although this claim might seem trivial at first glance, Hume emphasizes that it answers "one of the most sublime questions in philosophy, viz., *that*

49. It is worth noting here that Hume does not say that only the beliefs belonging to the judgment are approved by philosophers; this would exclude beliefs of memory and the senses. He does not say that all beliefs resulting from unreasoning imagination are whimsies, prejudices, or disapproved by philosophers.

concerning the power and efficacy of causes; where all the sciences seem so much interested" (T 1.3.14.2). If fully grasped, he expects that his answer will shock readers: "I am sensible, that of all the paradoxes, which I have had, or shall hereafter have occasion to advance in the course of this treatise, the present one is the most violent" (T 1.3.14.24). The shock lies in the fact that "necessary connexion" is not a feature of objects, but a purely subjective feeling that humans have about objects. It is an impression of reflection, not of sensation (T 1.3.14.22).

Since he clearly regards his theory as both important and controversial, he spends considerable time refuting rival views of necessary connection, recapping his arguments for his own view, and spelling out the implications. The central problem with every other account of necessary connection is that it violates the Copy Principle. "If we pretend, therefore, to have any just idea of this efficacy, we must produce some instance, wherein the efficacy is plainly discoverable to the mind, and its operations obvious to our consciousness or sensation" (T 1.3.14.6). But there is no instance where efficacy is discoverable to the mind, aside from the feeling we have when we make a causal inference.

Hume's account of the idea of necessity leads him to give two definitions of the causal relation. First, he defines causation as a philosophical relation. This definition includes only the objective characteristics of those items we think of as causes: "An object precedent and contiguous to another, and where all the objects resembling the former are plac'd in like relations of precedency and contiguity to those objects, that resemble the latter" (T 1.3.14.31). This definition does not include necessary connection (or synonyms like power, efficacy, or force) at all. He recognizes that this definition might "be esteem'd defective, because drawn from objects foreign to the cause." A mere constant conjunction of contiguous, successive items seems to lack the essential feature of the causal relation, that which distinguishes it from mere correlation (cf. T 1.3.2.11). Second, he defines causation as a "natural relation." In this sense, "A CAUSE is an object precedent and contiguous to another, and so united with it, that the idea of the one determines the mind to form the idea of the other, and the impression of the one to form a more lively idea of the other" (T 1.3.14.31). This definition includes the necessary connection, but only a deflationary, subjective necessary connection. The determination of the mind is the only necessary connection between the ideas of the causal relata.

Hume distinguishes between these two perspectives on causation from the beginning of Book 1. He notes early on that causation is both a natural relation among perceptions in the imagination (T 1.1.4.2) and a philosophical relation (T 1.1.5.9). He also notes the distinct roles that causation plays in inference qua philosophical and qua natural relation: "Thus tho' causation be a *philosophical* relation, as implying contiguity, succession, and constant conjunction, yet 'tis only so far as it is a *natural* relation, and produces an union among our ideas, that we are able to reason upon it, or draw any inference from it" (T 1.3.6.16). That is, experience of the philosophical relation of causation does not, by itself, produce an inference. However, due to the principle of custom, experience of the philosophical relation produces a natural relation between the ideas of the causal relata in the imagination. The natural relation—the determination of the mind—then produces causal inferences.

Although Hume's critical rejection of objective necessary connection is shocking, he does not construe it as a skeptical problem. He portrays his own constructive account of subjective necessary connection as satisfactory for all the purposes of common life and science. His treatment of objective necessary connection parallels his treatment of substances (T 1.1.6). It may be surprising to discover, also by means of the Copy Principle, that substances are just bundles of separable qualities. But neither the belief in substrata and accidents, nor the belief in objective necessary connections, is a core belief.

OVERVIEW OF PART 4

Whereas Parts 1–3 contain the large majority of Hume's constructive work in Book 1, Part 4 largely consists of destructive arguments.[50] In Section 1, Hume gives a skeptical argument against reason itself. In Section 2, he gives skeptical arguments against sensory beliefs; Section 4 completes this skeptical line of argument. Section 7 brings together these skeptical arguments against reason and sensory beliefs and develops a response to them. The next three chapters examine these sections in depth.

The remaining sections of Part 4 all argue against different forms of the substance metaphysic Hume already criticizes in *Treatise* 1.1.6. *Treatise* 1.4.3 critiques the view of that physical bodies consist of accidents inhering in substrata, *Treatise* 1.4.5 critiques the view that minds are either material or immaterial substances in this sense, and *Treatise* 1.4.6 critiques the view of the self as a simple substance remaining identical through change. Hume uses the Copy Principle to argue that an unperceived substratum is unintelligible in all these cases. He uses the Separability Principle to argue that no qualities are inseparable from any others. The constructive material he develops along the way in Part 4 serves essentially destructive ends.

Book 1 of the *Treatise* describes the various faculties and powers of the mind and how they work together to produce beliefs. It gives special attention

50. Ainslie's *Hume's True Scepticism* provides a searching examination of Part 4 in its entirety.

to how we form beliefs about unobserved, unremembered matters of fact on the basis of experience through the process of causal inference. It also raises questions about the epistemic status of these faculties, powers, and processes of the mind, and about the beliefs they produce. It is to these epistemic questions that I now turn.

Chapter Two

Philosophy and Justification in the *Treatise*

Debate about Hume's epistemology and epistemic skepticism labors under the difficulty that he does not use the term "epistemology." The only way to determine Hume's epistemology is to identify in his work a notion that bears a family resemblance to the cluster of concepts that contemporary philosophers talk about when they talk about "epistemic justification." The favorite notion is then christened as his basic epistemic idea or principle or concept or framework. Commentators have ascribed a range of fundamental epistemic ideas to Hume: reliability,[1] truth and probable truth,[2] coherence,[3] extensiveness and constancy,[4] stability,[5] long-term consensus,[6] reflexive self-confirmation,[7] reason,[8] agreeability and utility,[9] and more. There are good

1. Schmitt attributes to Hume a reliability account of justified belief, which he locates in a broader "veritistic epistemology, according to which true belief is the chief non-instrumental cognitive value and the primary value in terms of which knowledge is to be understood." Schmitt, *Hume's Epistemology*, 1.

2. Garrett argues that truth and probable truth are Hume's fundamental normative concepts of the epistemic domain. Garrett, *Hume*, 52–164.

3. Marie A. Martin proposes that, over against the traditional epistemic criterion of truth-conduciveness, Hume's criterion of correct epistemic principles is whether they result in "orderly and coherent judgments." Martin, "The Rational Warrant for Hume's General Rules," *Journal of the History of Philosophy*, 31, no. 2 (April 1993): 253–54.

4. Jack C. Lyons says that Hume and "the philosophers" endorse norms which license the influence of general rules which satisfy what Lyons calls the Extensiveness Constraint and the Constancy Constraint. Lyons, "General Rules," 247–78.

5. Loeb holds that for Hume, a belief is justified if and only if it results from a mechanism that "tends to produce stability in belief." Loeb, *Stability*, 13.

6. Michael Williams holds that "Hume's proposal is that long-term consensus replace truth as the goal of inquiry." Williams, "Hume's Skepticism," *The Oxford Handbook of Skepticism*, ed. John Greco (Oxford: Oxford University Press, 2008): 94.

7. William Edward Morris, "Belief, Probability, Normativity," *The Blackwell Guide to Hume's Treatise*, ed. Saul Traiger (Malden, MA: Blackwell Publishing, 2006): 77–94.

8. Millican, "Hume's 'Scepticism,'" 71–87.

9. Ridge, "Epistemology Moralized."

textual reasons to think that Hume cares about most or all of these concepts, and that they play a role in how he evaluates beliefs. However, there is always a textual leap involved in determining what role these concepts play in his "epistemology"—since he does not talk about it as such.

In this chapter, I offer an alternative approach to this debate about Hume's epistemology. Instead of looking for a central epistemic concept, I identify the terminology that Hume uses to evaluate beliefs and belief-forming practices. First, I argue that Hume's terminology of "philosophy" is the textual key to identifying his evaluations of beliefs from that standpoint which is normative for the sciences. Second, instead of looking for principles that cohere with a central epistemic concept, I gather the direct statements that Hume makes about the demands of philosophy. His most general statements about the standards of philosophical justification occur in one key passage, *Treatise* 1.4.4.1–2. This passage, taken together with others, indicates that in philosophy, a belief is defeasibly justified if and only if it is produced by a PIU (permanent, irresistible, and universal) principle of the imagination; it is justified overall if and only if it is defeasibly justified and faces no defeaters. I then identify those propensities that are permanent, irresistible, and universal. The result is a theory of philosophical justification that, as I argue in further chapters, coheres with the conclusions he draws from his skeptical arguments.

Although most commentators agree that *Treatise* 1.4.4.1–2 is an important passage, they give different answers to three key questions. First, does Hume endorse the standards of "philosophy," at least provisionally? I argue, along with the majority of commentators, that Hume does endorse them.[10] Second, are the standards of *Treatise* 1.4.4.1–2 Hume's *epistemic* standards? Some commentators deny that Hume's endorsement is epistemic.[11] However, a large number of commentators agree that, at least provisionally, Hume endorses these standards as an epistemic framework.[12] I argue that we do not

10. Morris denies that Hume endorses the principles of the "philosophers" either in *Treatise* 1.4.4.1–2 or anywhere else. Morris, "Hume's Conclusion," 99–100. Garrett denies that Hume endorses the distinctions made in *Treatise* 1.4.4.1–2. Garrett, *Cognition*, 229.

11. Fogelin does not seem to take the passage as drawing an epistemic distinction. Fogelin, *Hume's Skepticism*, 89–92. Oliver Johnson also seems to think that in the passage, Hume is drawing a distinction, on pragmatic grounds, within the class of epistemologically unjustified beliefs. Johnson, *The Mind of David Hume: A Companion to Book I of* A Treatise of Human Nature (Urbana and Chicago: University of Illinois Press, 1995), 274–76.

12. P. J. E. Kail, citing *Treatise* 1.4.4.1, also agrees that the products of permanent, irresistible, and universal propensities are epistemologically justified. P. J. E. Kail, *Projection and Realism in Hume's Philosophy* (New York: Oxford University Press, 2007), 68. Schmitt views the passage as Hume's criterion of defeasibly justifying operations. Schmitt, *Hume's Epistemology*, 286–312. Millican thinks that it is precisely the epistemic framework he endorses in T 1.4.4.1–2 that drives Hume to skepticism in T 1.4.7. Millican, "Hume's Chief Argument," 96–98. Noonan, Loeb, and Qu all argue in different ways that Hume's skeptical problems later force him to reject the epistemic principles expressed here in T 1.4.4.1–2. Noonan, *Hume on Knowledge*, 128–30; Loeb, *Stability*, 15, 154–62; Qu 115, 121–22, 129, 145.

need to ascribe an epistemology to Hume at all, but that his notion of "philosophy" is the best candidate for that role. Third, does Hume ever change his standards of philosophy under the pressure of skeptical challenges? I address this question in succeeding chapters, especially chapter 5.

"PHILOSOPHY" AND THE EVALUATION OF BELIEF

In this section, I survey Hume's usage of the term "philosophy" and its cognates in the *Treatise* (primarily Book 1) in order to determine its meaning. I find that Hume uses "philosophy" to refer to a normative method for pursuing the special sciences. I also note several of the rules which belong to this normative method. Although I emphasize their application in the sciences, I see no reason to think that Hume applies a different set of norms to belief-formation in the context of common life. The central norms of philosophy prescribe basing beliefs on experience and making good inductive inferences. Hume applies these standards to beliefs formed outside of the study as well as inside of it.

Humean "philosophy," as I construe it, is consistent with contemporaneous conceptions of philosophy. In the seventeenth and eighteenth centuries, there was no sharp distinction between philosophy and science.[13] *Scientia* referred to a reasoned, systematic body of knowledge, which was also the goal of philosophy. Thus, the two terms were often used interchangeably. Special sciences such as physics and chemistry were placed under the category of natural philosophy. Philosophy was not distinguished by its subject matter, which was expansive enough to include God, humanity, and the natural world, but by its rational, systematic method of inquiry. James Harris writes that in the eighteenth century, "To be a philosopher . . . was to approach a subject, any subject, in a careful, analytical, and inductive manner, and to derive from one's inquiries maximally general explanatory principles."[14] For example, John Locke writes that "Philosophy . . . is nothing but the true Knowledge of Things."[15] It was possible however for the actual practice of

13. See See James Harris, "Introduction," in *The Oxford Handbook to British Philosophy in the Eighteenth-Century*, edited by James A. Harris (New York: Oxford University Press, 2013), 6–9; Donald Rutherford, "Innovation and orthodoxy in early modern philosophy," in *The Cambridge Companion to Early Modern Philosophy*, edited by Donald Rutherford (New York: Cambridge University Press, 2006), 11–13; Werner Schneiders, "Concepts of Philosophy," in *The Cambridge History of Eighteenth-Century Philosophy*, edited by Knud Haakonssen (New York: Cambridge University Press, 2006), 28–33; Richard Serjeantson, "Becoming a Philosopher in Seventeenth-Century Britain," in *The Oxford Handbook of Seventeenth-Century British Philosophy*, edited by Peter R. Anstey (New York: Oxford University Press, 2013), 23–28.

14. Harris, *Hume*, 19; cf. 18–24.

15. John Locke, *An Essay concerning Human Understanding*, edited by Peter H. Nidditch (Oxford: Clarendon Press, 1975), 10.

philosophy to fall short of its normative definition. Locke laments that the "frivolous use of uncouth, affected, or unintelligible Terms" in "the Sciences" has made philosophy unwelcome in "well-bred Company, and polite Conversation." He aspires to remove "some of the Rubbish" that bad philosophy has put "in the way to Knowledge"—thus helping philosophy live up to its own normative ideal.

Throughout the *Treatise*, Hume refers to "philosophy" and "the philosophers" in two distinct ways: philosophy in principle and philosophy in actual practice. The distinction between philosophy de jure and philosophy de facto is apparent in the opening paragraph of the *Treatise*, where he makes virtually the same points as Locke in the passage I quoted in the preceding paragraph. Hume notes that the obvious defects of "the systems of the most eminent philosophers . . . seem to have drawn disgrace upon philosophy itself" (T Intro. 1). This statement distinguishes between philosophy de facto (as practiced defectively by even "the most eminent philosophers") and philosophy de jure ("philosophy itself"). His remarks about actual philosophers and their theories can be laudatory, neutral, or critical. On one hand, for example, he refers to George Berkeley and to John Locke as each "a great philosopher" in the course of appropriating some of their insights (T 1.1.7.1, 1.2.3.7). On the other hand, he criticizes ancient and modern philosophers (T 1.4.3–4). But to speak positively or critically of particular philosophers or bits of philosophy is entirely different from discussing "philosophy itself," philosophy de jure. When Hume talks about "the philosophers" and "philosophy" in principle, he always aligns himself with it and endorses it as dictating the proper way to conduct science.

To begin with, Hume's very pursuit of the "science of man" entails his endorsement of—his submission to—philosophy as a normative method. Hume can describe the project of the *Treatise* either as "An attempt to introduce the experimental Method of Reasoning into Moral Subjects" (subtitle) or as "the application of experimental philosophy to moral subjects" (T Intro. 7;). If these two descriptions are synonymous, then "philosophy" is a "method of reasoning." Furthermore, the relationship of philosophy to the particular arts and sciences is like the relationship of a king to his subjects: its "sovereign authority ought every where to be acknowledg'd" (T 1.4.5.34). Hume couples "philosophy and the sciences" in the first sentence of the *Treatise* (T Intro. 1). Just as he can speak of "science" in the singular as well as of its particular branches, "the sciences," (for example, the science of man), so also Hume can speak of "philosophy" and its particular branches, such as natural philosophy and moral philosophy.[16] He can variously describe his project both as a branch of science ("the science of man") and as a branch of

16. Cf. Garrett, *Cognition*, 4.

philosophy (specifically, "moral philosophy") (T Intro. 10). By contrast, he emphasizes that he is not engaged in natural philosophy (T 1.1.2.1, 1.2.5.4, 1.3.8.8). He refers to the pursuit of the special sciences as "philosophical researches" (T Intro. 6). "Philosophy" and "science" are often coextensive if not synonymous terms for Hume. "Philosophy" in general refers to a method of scientific inquiry governed by a set of normative principles. The particular branches of philosophy (or science) apply this method to different subject matter. Insofar as he participates in the sciences at all, Hume acknowledges the "sovereign authority" of philosophy, and endorses its norms.

Hume does not always use the word "philosophy" to directly denote a method. For example, when he talks about "the philosophy I am going to unfold" (T Intro. 3), he seems mean "the body of doctrine I am about to unfold." However, the methodological significance of "philosophy" seems to be the most basic, and to control other usages of the term. So, for example "the philosophy [body of doctrine] that I am about to unfold" is a body of beliefs arrived at by way of the methodological norms of philosophy.

Hume never retracts his endorsement of philosophy as normative for the sciences. Philosophy is not one method among others for pursuing the special sciences; it is the only proper method. The only alternative method of positive belief-formation and theory construction that Hume mentions is superstition (T 1.4.7.13), and this he categorically rejects. As I discuss later, Hume does qualify his endorsement of philosophy in one sense. We are sometimes practically justified in ignoring the demands of philosophy. But insofar as we do deviate from philosophy, we are no longer engaged in the proper conduct of science.

Throughout Book 1 of the *Treatise* Hume indicates many normative principles that govern philosophy. These principles all govern our doxastic activities, those activities related to belief-formation. My aim in what follows is to draw attention to the many important texts where these doxastic norms are directly described in terms of "philosophy." These passages corroborate my claim that "philosophy" is the terminological key to identifying Hume's normative judgments in the context of scientific inquiry. I do not however aim to catalog every such norm that Hume implicitly or explicitly endorses.

As we have seen, "philosophy" and its methodological norms are in view from the first sentence of the "Introduction." In this context Hume lays down the authority of observation and rules out the postulation of objective necessary connections. He also says that to accept propositions without sufficient evidence, to deduce consequences invalidly, or to espouse incoherent theories all draw "disgrace upon philosophy," presumably because these actions violate its norms (T Intro. 1).

The faculty of reason is essential to philosophy. Philosophy brings questions "before the tribunal of human reason" (T Intro. 1). To reject all refined and elaborate reasoning is to "cut off entirely all science and philosophy" (T 1.4.7.7). Philosophers disapprove of education (that is, inculcation of belief by brute repetition) because "its maxims are frequently contrary to reason" (T 1.3.9.19). *Treatise* 1.3.13, "Of unphilosophical probability," describes four kinds of probabilistic belief-formation processes that do not obtain the sanction of the philosophers (T 1.3.13.1). These unphilosophical forms of probability are contrasted with the probabilistic belief-forming processes which "are receiv'd by philosophers, and allow'd to be reasonable foundations of belief and opinion" (T 1.3.13.1). These approved forms of probability are those "which are deriv'd from an *imperfect* experience and from *contrary* causes," as well as probability arising from analogy (T 1.3.12.25). The "Rules by which to judge of causes and effects" (T 1.3.15) also belong to Hume's "philosophy." The fourth rule "is the source of most of our philosophical reasonings" (T 1.3.15.4). In contrast to the baroque systems of "Our scholastic head-pieces and logicians," Hume gives this compact set of rules "to direct our judgment, in philosophy," both moral and natural (T 1.3.15.11).

In a passage of programmatic significance, Hume distinguishes between principles of the imagination which "are receiv'd by philosophy" and those which are "rejected" by philosophy" (T 1.4.4.1). Philosophy approves of those principles of the imagination which are "permanent, irresistible, and universal," and disapproves of those which are "changeable, weak, and irregular" (T 1.4.4.1). In the next paragraph Hume characterizes philosophically approved principles as "solid, permanent, and consistent" (T 1.4.4.2). He gives as an example of an approved principle "the customary transition from causes to effects, and from effects to causes" (T 1.4.4.1). He gives as an example of a "trivial propensity of the imagination" which philosophy condemns the propensity to project internal emotions on external objects (T 1.4.3.11).

Hume identifies several more or less miscellaneous norms of philosophy simply in passing. It is the duty of philosophers to clarify obscure ideas: "If its [an idea's] weakness render it obscure, 'tis our business to remedy that defect, as much as possible, by keeping the idea steady and precise; and till we have done so, 'tis in vain to pretend to reasoning and philosophy" (T 1.3.1.7). Philosophers ought to define their terms as clearly and precisely as possible—but no more than is possible (T 1.3.7.7).[17] Philosophers should be slow to accept new hypotheses: "A scrupulous hesitation to receive any new hypothesis is

17. Hume admits that the variety of terms whereby he defines the manner of conceiving those ideas which constitute beliefs may seem "unphilosophical," but in fact he counters that "in philosophy we can go no farther, than assert, that it is something *felt* by the mind, which distinguishes the ideas of the judgment from the fictions of the imagination" (T 1.3.7.7). Here Hume is concerned to show that he does in fact abide by the norms of philosophy, even if at first it may appear otherwise. The particular norm in view is that we ought to give clear and precise definitions to our terms.

so laudable a disposition in philosophers, and so necessary to the examination of truth, that it deserves to be comply'd with, and requires that every argument be produc'd, which may tend to their satisfaction, and every objection remov'd, which may stop them in their reasoning" (T 1.3.9.1). False philosophy predicates qualities of objects with which those qualities are incompatible: "But when, instead of meaning these unknown qualities, we make the terms of power and efficacy signify something, of which we have a clear idea, and which is incompatible with those objects, to which we apply it, obscurity and error begin then to take place, and we are led astray by a false philosophy" (T 1.3.14.27). True philosophy involves correcting our propensity to conceptual confusion: "Our propensity to this mistake is so great from the resemblance above-mention'd, that we fall into it before we are aware; and tho' we incessantly correct ourselves by reflection, and return to a more accurate method of thinking, yet we cannot long sustain our philosophy, or take off this biass from the imagination" (T 1.4.6.6). True philosophers do not successively assent to contradictory principles: "Or in case we prefer neither of them, but successively assent to both, as is usual among philosophers, with what confidence can we afterwards usurp that glorious title, when we thus knowingly embrace a manifest contradiction?" (T 1.4.7.4).

The norms of philosophy are distinct from prudential and moral norms. An action is prudent if it advances our own long-term self-interest. Morally, Hume approves of qualities which are immediately agreeable or useful to their possessors or to others (T 3.3.1.30, EPM 9.1). Philosophy does not evaluate beliefs or belief-forming processes on the basis of their agreeability or utility, either for oneself or others. The belief-forming processes sanctioned by philosophy are, generally speaking, indispensably useful to human life (T 1.4.4.1). But there are exceptions to this general statement. As I will show later when I discuss Hume's skeptical challenges, following philosophy consistently leads to a suspension of our practically indispensable beliefs. In any case, just because philosophically approved belief-forming propensities are useful does not entail that these propensities are philosophically approved *because* they are useful. In fact, as Qu points out,[18] Hume says that in "philosophical debates" it is highly "blameable" to reject a claim because of its allegedly "dangerous," immoral, or irreligious consequences:

> There is no method of reasoning more common, and yet none more blameable, than in philosophical debates to endeavour to refute any hypothesis by a pretext of its dangerous consequences to religion and morality. When any opinion leads us into absurdities, 'tis certainly false; but 'tis not certain an opinion is false, because 'tis of dangerous consequence. (T 2.3.2.3)

18. Qu, "Hume's Practically Epistemic Conclusions?," 509

Qu also notes Hume's similar remark that "While a warm imagination is allow'd to enter into philosophy, and hypotheses embrac'd merely for being specious and agreeable, we can never have any steady principles, nor any sentiments, which will suit with common practice and experience" (T 1.4.7.14). These texts prove that philosophical justification is not, for Hume, dependent upon agreeability and utility.

In my view, we can treat Hume's theory of philosophical justification as his theory of epistemic justification, and that is what I will do in the remainder of this book. If a belief or a belief-forming propensity receives the sanction of philosophy, then it is philosophically justified. Hume treats these beliefs, and only these beliefs, as well-grounded, warranted, intellectually respectable both in the domain of science and in common life. This description makes "philosophical" justification sound very similar to "epistemic" justification.

However, I make two caveats about identifying "philosophical" and "epistemic" justification. First, we can make perfectly good sense out of Hume's text relying only on his own term ("philosophy") without discussing "epistemology" at all. Talk of his "epistemology" is optional. I talk about Hume's epistemology because it makes it easier to engage with other commentators who do so, and because it facilitates fruitful comparison and contrast of Hume with contemporary philosophy. Some readers may think that the differences between Humean "philosophy" and epistemology (as they construe it) are so great that the two concepts cannot be identified. To such readers I suggest that if Hume's "philosophy" is not his epistemology, then it is what he has instead of an epistemology. In that case, we had better dispense with talking about his epistemology at all and limit ourselves to his "philosophy" when we want to talk about the intellectual acceptability of beliefs.

Secondly, it is important that we not automatically transfer contemporary assumptions about epistemic justification into Hume's concept of philosophy. For example, contemporary epistemologists often think of justified beliefs as essentially reliable or truth-conducive, or as fulfilling a kind of epistemic duty. Say that we as interpreters share these assumptions about what epistemic justification essentially involves. We might then conclude, without further ado, that philosophical justification for Hume is truth-conducive or deontological. Such a conclusion would be rash, anachronistic, and unwarranted without clear textual grounds.

As for truth-conduciveness, it is plausible to think that philosophy sanctions those doxastic practices which are most likely to lead to truth. In the section "Of curiosity, or the love of truth," Hume compares philosophy to hunting, and truth to the hunter's quarry (T 2.3.10.8–9). The primary point of the comparison is to show how philosophy, like hunting, affords the pleasure of pursuit. But the analogy shows something else about the relationship between philosophy and truth: philosophy is the most likely method for obtaining truth, just as hunting is the most likely method for catching prey. The whole section (T 2.3.10) presupposes that the way to try to gratify the passion of curiosity (the love of truth) is to pursue reason and philosophy.

However, just because philosophically sanctioned doxastic practices are *more* likely to lead to truth than the alternatives, does not mean that they are in fact truth-conducive. On my reading, following philosophy does not necessarily or even probably lead to truth, frequently or at all.[19] Hume opens the *Treatise* by expressing diffidence about whether truth is in human reach: "For *if* truth be at all within the reach of human capacity, 'tis certain it must lie very deep and abstruse; and to hope we shall arrive at it without pains, while the greatest geniuses have failed with the utmost pains, must certainly be esteemed sufficiently vain and presumptuous" (T Intro. 3, italics mine). He eventually concludes that consistent adherence to philosophy does not lead to true or probably true beliefs, but to suspended judgment: "Philosophy wou'd render us entirely *Pyrrhonian,* were not nature too strong for it" (AT 27). As I will discuss later, if Hume's criterion for philosophical norms were truth-conduciveness, and he found that his norms rendered him Pyrrhonian, he would have to reject his norms as philosophically unacceptable. As a matter of fact, when he does find that philosophical norms lead to global agnosticism, he does not reconsider the normative content of philosophy, but rather his commitment to philosophy (T 1.4.7). When he returns to philosophy, he recommends it over superstition not on grounds of its reliability, but because it is safer and more agreeable (T 1.4.7.13). At the close of Book I he expresses diffidence about the attainability of truth: "we might hope to establish a system or set of opinions, which if not true (for *that, perhaps, is too much to be hop'd for*) might at least be satisfactory to the human mind, and might stand the test of the most critical examination" (T 1.4.7.14, italics mine).

CONDITIONS FOR JUSTIFICATION

The distinction between PIU and CWI principles of the imagination in *Treatise* 1.4.4.1–2 provides the most general perspective on Hume's norms of philosophy. In the last chapter I described this distinction in relation to reason and the imagination. Now I want to explain its normative upshot. The passage comes between Hume's criticism of the substance metaphysic of the "antient" philosophers (T 1.4.3) and his criticism of the "modern" account of representative realism that relies on the primary and secondary quality

19. Owen also argues that Hume cannot claim truth-conduciveness for his philosophical norms. Owen, *Hume's Reason*, 207–8.

distinction (T 1.4.4). In order to sustain his criticisms of the ancient and modern philosophers, he needs to show how his reliance on the imagination (which is, "according to my own confession, . . . the ultimate judge of all systems of philosophy") is different than theirs (T 1.4.4.1). He needs to show that his reliance on imagination is philosophically justified in a way that the ancient and modern reliance on the imagination is not. In that context, he draws a line between two classes of principles of the imagination, one of which is "received by philosophy" and the other is "rejected" by philosophy:

> In order to justify myself, I must distinguish in the imagination betwixt the principles which are permanent, irresistable, and universal; such as the customary transition from causes to effects, and from effects to causes: And the principles, which are changeable, weak, and irregular; such as those I have just now taken notice of. The former are the foundation of all our thoughts and actions, so that upon their removal human nature must immediately perish and go to ruin. The latter are neither unavoidable to mankind, nor necessary, or so much as useful in the conduct of life; but on the contrary are observ'd only to take place in weak minds, and being opposite to the other principles of custom and reasoning, may easily be subverted by a due contrast and opposition. For this reason the former are received by philosophy, and the latter rejected. (T 1.4.4.1)

I base my account of philosophical justification mainly on this passage, which I will now unpack.

The PIU principles are natural in the sense that they are part of what constitutes human nature as such. Hume describes PIU principles as "the foundation of all our thoughts and actions, so that upon their removal human nature must immediately perish and go to ruin" (T 1.4.4.1). Just causal inference is a paradigmatic operation of PIU principles. Thus, "One who concludes somebody to be near him, when he hears an articulate voice in the dark, reasons justly and naturally" (T 1.4.4.1). These principles are "the solid, permanent, and consistent principles of the imagination" (T 1.4.4.2).

CWI principles are not natural in the same way as the PIU principles. They are not "unavoidable . . . but on the contrary are observ'd only to take place in weak minds, and being opposite to the other principles of custom and reasoning, may be easily subverted by a due contrast and opposition" (T 1.4.4.1). These principles, "however common, are neither universal nor unavoidable in human nature" (T 1.4.4.2). CWI principles are "natural," but only in the broad sense in which the cause of a sickness is natural (T 1.4.4.1). Later in the *Treatise* Hume writes that "nature" may be opposed to miracles (T 3.1.2.7), to that which is "rare and unusual" (T 3.1.2.8), or to artifice (T 3.1.2.9). CWI principles and the causes of maladies are neither miraculous nor artificial nor rare, so they are "natural" in all these senses. He gives as an example of a "trivial propensity of the imagination" which philosophy condemns the propensity to project internal emotions on external objects (T 1.4.3.11).

It might seem as though philosophical approval depends upon the utility of the principles of the imagination. Hume emphasizes that upon the removal of PIU principles, "human nature must immediately perish and go to ruin" (T 1.4.4.1). That is, they are necessary and useful for human life. CWI principles are "neither . . . necessary, nor so much as useful in the conduct of life." They are like a malady, "contrary to health, the most agreeable and most natural situation of man" (T 1.4.4.1). After noting the relative of utility of the two sets of principles, Hume writes that "*For this reason* the former are receiv'd by philosophy, and the latter rejected" (italics mine). This sounds like philosophical approval is based on relative practical benefit.

However, there are stronger reasons for thinking that the practical benefits of PIU principles are incidental to the approval of philosophy. First, "For this reason" need not refer back to all the reasons in the preceding descriptions of PIU and CWI principles (which include descriptions of their relative utility). The immediately preceding "reason" is that CWI principles "may easily be subverted by a due contrast and opposition" with "the other principles of custom and reasoning" (T 1.4.4.1). The most natural reading of "For this reason" is that philosophy receives PIU principles and rejects CWI principles because the former subvert the latter when they are duly contrasted.

Second, as noted earlier, philosophy does not accept or reject beliefs based on their agreeability or utility (T 2.3.2.3; T 1.4.7.14). This provides at least some reason to doubt that philosophy would accept or reject belief-forming propensities based on whether they produce useful beliefs. Third, although Hume mentions the relative utility of these principles in close connection with their other properties, he gives more emphasis to their other properties. When he briefly recapitulates the normative contrast between approved and unapproved principles in the next paragraph, he does not mention practical benefit. He says only that the "defect" of the disapproved principles is that they "are neither universal nor unavoidable in human nature" (T 1.4.4.2). The correlative virtue of the approved principles is that they are "solid, permanent, and consistent." Neither agreeability nor utility make an appearance.

Fourth, adhering to PIU principles with absolute consistency results in suspension of virtually all beliefs, a condition that is as self-destructive as any. In the final section of Book 1, Hume refers back to the skeptical argument against reason that he makes in *Treatise* 1.4.1 (T 1.4.7.7n53). With reference specifically to that argument, Hume writes that "a resolution to reject all the trivial suggestions of the fancy, and adhere to the understanding, that is, to the general and more establish'd properties of the imagination; even this resolution, if steadily executed, wou'd be dangerous, and attended with the most fatal consequences" (T 1.4.7.7). By "trivial suggestions of the fancy," Hume means CWI principles, and by "the general and more establish'd properties of the imagination," he means the PIU principles. His point is that if we adhere to PIU principles in our response to the skeptical argument against reason, the results will be fatal.[20] So PIU principles are not always more useful than CWI principles. Granted, this passage comes long after his initial explanation of the PIU/CWI distinction. But he has already made the key argument on which that passage is based in *Treatise* 1.4.1. There is no reason to doubt that Hume holds a consistent view about the utility of the PIU principles throughout Part 4 of Book 1. Based on these four considerations, I conclude that adherence to PIU principles is usually but not always useful, and that adherence to CWI principles is usually but not always harmful. Philosophy as such values PIU principles regardless of their utility.

According to *Treatise* 1.4.4.1–2, philosophy primarily approves and rejects principles of the imagination. Philosophy approves and rejects individual beliefs, like causal inferences or the projections of the ancient philosophers, only on account of the principles that produce them. This gives Hume a propensity-based epistemology, wherein the epistemic status of a belief derives from the properties of the propensity that produces it.[21] Derivatively, then, philosophy approves of beliefs that PIU principles produce, and disapproves of beliefs that CWI principles produce. Hume gives examples of beliefs approved by philosophy because of the propensity that produces it: "One who concludes somebody to be near him, when he hears an articulate voice in the dark, reasons justly and naturally; tho' that conclusion be deriv'd from nothing but custom, which infixes and enlivens the idea of a human creature, on account of his usual conjunction with the present impression" (T 1.4.4.1). A contrasting example shows how philosophy disapproves of particular beliefs on account of its source propensity. One "who is tormented he knows not why, with the apprehension of spectres in the dark," has beliefs "arising from natural causes" of the wrong sort (T 1.4.4.1). By the same token, if it turns out that CWI principles produce the modern philosophy, then philosophy must reject it (T 1.4.4.2).

The philosophical justification of beliefs produced by PIU principles is defeasible. In Hume's epistemology—and arguably in any viable epistemology—there is a distinction between prima facie, defeasible justification and

20. Millican, "Hume's Chief Argument," 96–97; Qu, *Hume's Epistemological Evolution*, 118–23.
21. In this respect I agree with Loeb and Schmitt. Loeb, *Stability*, 12–13; Schmitt, *Hume's Epistemology*, 28–29.

overall justification, or justification *tout court, ultima facie*.[22] Obviously Hume does not use any of these contemporary terms of art. However, the way he treats the philosophical status of various beliefs indicates that he implicitly works with these concepts or something very near them. For example, Hume would surely agree that when new evidence emerges, the philosophical acceptability of a belief may change. PIU principles may lead us to infer, on the basis of observational evidence, that all swans are white. So far forth, this belief is received by philosophers. But if and when we observe a black swan, our previous belief is no longer justified. The belief that all swans are white possessed defeasible justification but not overall justification. Hume discusses cases like this when he takes up prejudice (T 1.3.13.7–8). PIU principles may lead us to believe some general rule, such as that "An *Irishman* cannot have wit, and a *Frenchman* cannot have solidity" (T 1.3.13.7). When PIU principles produce belief in a general rule, the belief possesses defeasible justification. But if we hold these beliefs "tho' the conversation of the former in any instance be visibly very agreeable, and of the latter very judicious," then the belief becomes a prejudice, a form of "unphilosophical probability." In such cases, "men form general rules, and allow them to influence their judgment, even contrary to present observation and experience" (T 1.3.13.8). This analysis reflects Hume's assumption that though PIU principles may produce a philosophically justified belief, such beliefs are no longer philosophically justified when observation contradicts them.

I can now give a general account of the beliefs that philosophy approves and rejects. From the perspective of "philosophy" (that is, according to Hume's epistemology), *a belief is defeasibly justified if and only if it is produced by a PIU principle of the imagination; a belief is justified overall if and only if it is defeasibly justified and faces no defeaters*. This account merely formalizes two textual observations. The first observation is that philosophy approves of beliefs produced by PIU principles, and not those produced by CWI principles. The second observation is that philosophy can cease approving of those beliefs when new evidence emerges. Taken together, these two observations yield the simple but powerful Humean theory of epistemic justification in the *Treatise*. In following chapters, I use this account to determine the epistemic status of core beliefs and the epistemic force of the various skeptical arguments that Hume presents.

---

22. Lyons helpfully discusses defeasibility and prima facie versus *ultima facie* justification in Hume's epistemology. Lyons, "General Rules," 276n30. Schmitt makes heavy usage of defeasible versus overall justification in *Hume's Epistemology*. He provides a helpful discussion of defeasible versus overall justification, both with respect to epistemology generally and to Hume in particular. Schmitt, *Hume's Epistemology*, 28–33. Schmitt adduces different textual grounds than I do for attributing to Hume the defeasible/overall distinction. Ibid. 29–30. But he rightly points out that "the distinction is essential equipment for a plausible epistemology" generally. Ibid. 30.

The *Treatise* indicates several of the belief-producing processes of which the philosophers approve, and that defeasibly justify their products. Hume relies on these processes throughout the constructive portions of the *Treatise*. Any of these processes can go wrong. For example, we sometimes make bad inferences, or misremember things. But philosophy approves of these processes when they go right.

First, the entire *Treatise* assumes the defeasible justification of introspective beliefs about our own perceptions. Hume appeals to this direct introspective awareness to establish the difference between impressions and ideas in the first paragraph of the body of the *Treatise* (T 1.1.1.1). We can form ideas of ideas and impressions (T 1.1.1.11), and beliefs in the existence of ideas and impressions (T 1.3.8.17). These second-order beliefs about our own perceptions must occur in the imagination. If they are philosophically approved beliefs, as at least some of them seem to be, then they must be produced by PIU principles.

Second, intuition and demonstrative reasoning afford knowledge (in the strict sense of certainty) of philosophical relations between ideas (T 1.3.1). In a passage cited earlier, Hume speaks of "the understanding, that is, . . . the general and more establish'd properties of the imagination" (T 1.4.7.7). If intuition and demonstrative reasoning belong to the understanding, as I argued in chapter 1, and the understanding belongs to the PIU principles, as this passage seems to indicate, then intuition and demonstrative reasoning must be PIU principles.[23] Even without a specific proof text stating that philosophy sanctions intuition and demonstrative reasoning, there can be no serious doubt that it does.

Third, probable reasoning is the prime example of a PIU principle of the imagination in *Treatise* 1.4.4.1: "such as the customary transition from causes to effects, and from effects to causes." Furthermore, probable reasoning belongs to the understanding and hence, as pointed out in the previous paragraph, to "the general and more establish'd properties of the imagination" (T 1.4.7.7).

Fourth, philosophers approve of memorial beliefs. Memory is a distinct faculty from the imagination, so it might initially seem like the PIU account does not determine its philosophical status. Hume however writes that "The memory" is "founded on the imagination, or the vivacity of our ideas." (T 1.4.7.3). That is, the vivifying principles of the imagination give memory ideas the liveliness that constitutes them as beliefs. So the PIU account does cover memory. Although memory, like any other faculty, can malfunction (T 1.3.5.6), Hume raises no specific skeptical argument against it.[24] Further-

23. Thanks to an anonymous referee for pointing this out.

24. In *Treatise* 1.4.7.3 he does seem to question its trustworthiness, along with all the other belief-producing processes of the imagination. I argue in chapter 5 that in this paragraph Hume is making a *reductio* argument whose premises he himself does not accept. So the argument does not entail that philosophy disapproves of these faculties and processes.

more, memory plays an essential role in causal inference (T 1.3.6.2), which Hume says is a process produced by PIU principles. If memory does not even defeasibly justify its beliefs, it seems Hume would have to call attention to the fact somehow. Every indication is that memorial beliefs are produced by PIU principles of the imagination.[25]

Fifth, PIU principles produce sensory beliefs in mind-independent objects (continued and distinct existences). I argue at greater length for the defeasible justification of sensory beliefs in chapter 4. In short, the imagination, rather than the senses in the strict sense, produce belief in body (T 1.4.2.14). But it is psychologically impossible to resist sensory beliefs for long (T 1.4.2.1, 57), which shows that they are produced by PIU principles.

One possible objection to my reading of *Treatise* 1.4.4.1–2 is that Hume would surely not wait until the end of Book 1 to state his most general philosophical norms. If what he says there really undergirds his entire scientific method and governs all the conclusions he has reached in the science of man thus far, he would have stated it much earlier. But by *Treatise* 1.4.4, Hume has already established all of his main positive conclusions about the understanding (T 1.1–3) and made his major skeptical arguments (T 1.4.1–2). He must have an epistemology in place long before *Treatise* 1.4.4.

I reply that *Treatise* 1.4.4.1–2 is a natural place for Hume to give his highest level of generalizations about the norms that his investigation has presupposed all along. As I documented earlier, Hume talks about the norms of his philosophical method from the very beginning of the *Treatise*. For example, from the "Introduction" onward he makes it clear that philosophy endorses reason. But during his investigations, Hume turns up a surprising fact: probable reasoning depends upon a principle of the unreasoning imagination (T 1.3.6). Only after making this discovery can Hume give an adequate characterization of those belief-forming processes that philosophy has approved all along: philosophy approves of PIU principles of the imagination, not CWI principles. This parallels his approach to moral norms: he first investigates the particular qualities of which we approve before formulating his general theory that these are all qualities that are agreeable or useful to ourselves or others. But it is not as though our approval of particular qualities depends upon an antecedent knowledge of the general norms. In fact, the reverse is true.

Another objection to my reading is that Hume would surely state his general epistemological norms in a section devoted to that topic, not as a digression from another topic. *Treatise* 1.4.4.1 cannot be a major statement of

25. If I am wrong, and memorial beliefs are not produced by the imagination at all, the only consequence is that my account of justification is not quite fully general, since there is at least one class of beliefs it does not cover.

Hume's epistemology, because it appears simply as a minor, off-hand clarification that he makes prior to destroying the "modern philosophy" of physical objects. If what he says here really sums up his view of philosophical norms, he would have drawn more direct attention to it for its own sake.

But I reply that Hume has no reason to highlight his epistemic norms for their own sake at this point in the *Treatise*. He understandably calls attention to his methodological commitments in the "Introduction." Even there, he assumes that his normative method is common knowledge. He is committed to the same method of "experimental philosophy" that had been pioneered by Francis Bacon and applied to the science of man by "some late philosophers of England" (T Intro.7). After that, he only needs to address philosophical norms directly if his discoveries in the science of man shed new light on them, or if a particular argument requires them. Both conditions are met in *Treatise* 1.4.4.1–2. The particular arguments about belief in body require him to clarify the approved and disapproved principles of the imagination, in light of his own discoveries about the imagination. The passage is important to interpreters for different reasons than it is to Hume. For Hume, it merely plays a minor supporting role in his critique of the modern philosophy. But for interpreters, it provides a tool for identifying the epistemic norms that Hume takes for granted all along.

I conclude then that *Treatise* 1.4.4.1–2 does provide the clearest and most direct generalized account of Hume's philosophical (that is, epistemic) norms. A wide range of interpreters can agree that Hume endorses this account; it does not presuppose any substantive claims about Hume's fundamental epistemic concept. The account ascribes to Hume a fairly uncontroversial and quotidian set of sources of justified belief: introspection, intuition, sensory experience, memory, and reason. With this minimalist, textually anchored account of justification in place, I can now assess the significance of skeptical arguments.

Chapter Three

Reason and Skepticism in the *Treatise*

Probably Hume's most famous contribution to philosophy is the so-called "problem of induction" that he develops in *Treatise* 1.3.6 and *Enquiry Concerning Human Understanding*, Section 4. Hume does not use the term "induction" and he does not refer to a "problem." He does argue that the faculty of reason does not determine us to infer the existence of an unobserved causal relatum from an observed relatum. These sorts of causal inferences are materially the same as what philosophers now call inductive inferences. The denial that causal inferences derive from reason might seem to deprive them of any epistemic justification. So it is reasonable to ask whether Hume thinks his account of causal inference raises a skeptical problem for induction.

In this chapter I argue that Hume does not intend *Treatise* 1.3.6 as a skeptical argument against induction. On the contrary, he holds that causal inference defeasibly justifies its products. Causal inference is produced by the permanent, irresistible, and universal principles of the imagination; Hume and the philosophers approve of these principles without reservation. Nonetheless, Hume is a radical epistemic skeptic about the conclusions of reason. Hume's real worry about reason (demonstrative as well as probable) is an undermining defeater argument that he mounts in *Treatise* 1.4.1.[1] I argue against non-skeptical interpreters that, according to Hume, the recursive argument against reason does epistemologically defeat every deliverance of reason.

---

1. De Pierris notes that the dominant view of recent interpreters is that Hume mounts no skeptical arguments prior to Part 4 of Book 1 of the *Treatise*. De Pierris, *Ideas, Evidence, and Method*, 277. De Pierris herself swims against this current, arguing that Hume does mount an inductive skeptical argument in T 1.3.6 which he continues to develop throughout Parts 3 and 4. Ibid., 259–306.

HUME'S ANALYSIS OF CAUSAL INFERENCE IN *TREATISE* 1.3.6

Treatise 1.3.6 ("Of the inference from the impression to the idea") concludes Hume's psychological investigation into the nature of causal inference. In *Treatise* 1.3.4, he lays it down that all causal inferences must begin with an impression or memory, and conclude to an unobserved, unremembered idea. In *Treatise* 1.3.5, he explains that a superior degree of force and vivacity distinguishes impressions and memories from ideas merely entertained in the imagination. In *Treatise* 1.3.6, he takes up the question of how we make a causal inference from an impression or memory to an idea that has sufficient force and vivacity to count as a belief.

He begins with some preliminary points. Since neither intuition nor demonstrative reasoning produce causal inference, it follows that "EXPERIENCE" does (T 1.3.6.1–2; cf. EHU 4.6–13). When we remember the constant conjunction of two species of objects in our past experience, "Without any farther ceremony, we call the one *cause* and the other *effect*, and infer the existence of the one from that of the other" (T 1.3.6.2). Constant conjunction, then, is another essential characteristic of the causal relation, in addition to contiguity and temporal succession (T 1.3.6.3).

Now Hume can pose the central question of the section: "Whether experience produces the idea by means of the understanding or imagination; whether we are determin'd by reason to make the transition, or by a certain association and relation of perceptions" (T 1.3.6.4).

Hume states the question twice in a row, varying his terms. The question is worth unpacking. First, note that the question is about *causal inference*—the production of the believed idea of a causal relatum, the transition from the impression to the idea after repeated experience. Furthermore, the question is about the *cause* of the causal inference: by what means is it produced, and what determines it to occur? There are two possible causes. One possible cause is "the understanding," "reason." The other possible cause is "the imagination," "a certain association and relation of perceptions." His answer to the question, the main thesis of the section, is that causal inferences are not produced by reason but by the imagination.

The main argument for this thesis begins at paragraph four and continues through the end of the section. In short, it runs as follows:

1. Causal inference is produced either by the understanding or by the imagination. (T 1.3.6.4).
2. Causal inference is not produced by the understanding. (T 1.3.6.5–11).

3. Therefore, causal inference is produced by the imagination. (T 1.3.6.12–16)

Hume takes it for granted without argument that one of two faculties, the understanding or the imagination, must produce causal inferences. He focuses rather on the controversial claim that the understanding does *not* produce causal inferences.

The sub-argument for the crucial second premise hinges upon the source of belief in what is often called the Uniformity Principle (UP). The UP is the name given to the principle "*that instances, of which we have had no experience, must resemble those, of which we have had experience, and that the course of nature continues always uniformly the same*" (T 1.3.6.4). The sub-argument runs as follows.

1. If causal inference is produced by the understanding, then belief in the UP is produced by the understanding. (T 1.3.6.4)
2. If belief in the UP is produced by the understanding, then it is produced either by demonstrative or probable reasoning. (T 1.3.6.4)[2]
3. Therefore, if causal inference is produced by the understanding, then belief in the UP is produced either by demonstrative or probable reasoning. (From 1, 2)
4. But belief in the UP is not produced by demonstrative reasoning (T 1.3.6.5),
5. And belief in the UP is not produced by probable reasoning. (T 1.3.6.6–7)
6. Therefore, belief in the UP is not produced either by demonstrative or probable reasoning. (From 4, 5).
7. Therefore, causal inference is not produced by the understanding. (From 3, 6)

Hume makes short work of the fourth premise, the claim that belief in the UP is not produced by demonstrative reasoning. In a piece of demonstrative reasoning, it is inconceivable that the premises should be true and the conclusion false. But "We can at least conceive a change in the course of nature; which sufficiently proves, that such a change is not absolutely impossible" (T 1.3.6.5). So belief in the UP is not produced by demonstrative reasoning.

2. "In order therefore to clear up this matter, let us consider all the arguments, upon which such a proposition may be suppos'd to be founded; and as these must be deriv'd either from *knowledge* or *probability*, let us cast our eye on each of these degrees of evidence, and see whether they afford any just conclusion of this nature" (T 1.3.6.4). Hume is obviously using "knowledge" in the strict sense that applies only to an object of intuition or demonstrative reasoning (T 1.3.1.1).

Hume's argument for the crucial fifth premise runs as follows (T 1.3.6.7):

1. All probable reasoning is founded on the supposition of the UP.[3]
2. If the supposition of the UP is founded on probable reasoning, then the UP is both the cause and effect of probable reasoning. (From 1)
3. "The same principle cannot be both the cause and effect of another."
4. Therefore, the supposition of the UP is not both the cause and effect of probable reasoning. (From 3)
5. Therefore, the supposition of the UP is not founded on probable reasoning.[4] (From 2, 4)

By the end of paragraph seven, Hume has completed his argument that causal inferences are not produced by reason. He then anticipates and refutes another possible objection (T 1.3.6.8–10). According to this objection, we can make rational arguments from past observation to future effects, without relying on the problematic UP. Hume states the argument twice in *Treatise* 1.3.6.8. I formalize it as follows, footnoting the textual basis for each lemma.

1. If an object has always been found to produce an effect, then it has the power to produce that effect.[5]
2. If an object has the power to produce an effect, then it always does produce the effect.[6]
3. Therefore, if an object has always been found to produce an effect, then it always produces the effect.[7]

If this argument succeeds, then reason does produce causal inferences after all.

Hume briefly alludes to some objections he might press against this argument, such as that "objective powers" are inconceivable (T 1.3.6.9, referring forward to 1.3.14). But his main objection is to Premise 1 (T 1.3.6.10). Even if past productions indicate the past powers of an object, we have no argument to show that the object continues to possess those same powers. We cannot directly observe an object's powers. Its powers might conceivably change

3. "probability is founded on the presumption of a resemblance betwixt those objects, of which we have had experience, and those, of which we have had none" (T 1.3.6.7). As Garrett notes, this supposition or presumption need not be, and often is not, an explicitly formulated belief. *Hume* 177.

4. "therefore 'tis impossible this presumption [of the UP] can arise from probability."

5. "Such an object is always found to produce another. 'Tis impossible it cou'd have this effect, if it was not endow'd with a power of production. . . . The past production implies a power:"

6. "The power necessarily implies the effect . . . The power implies a new production:"

7. "therefore there is a just foundation for drawing a conclusion from the existence of one object to that of its usual attendant. . . . And the new production is what we infer from the power and the past production."

through time, so there is no demonstrative argument for the continued possession of powers. Any probable argument that similar objects always have similar powers relies on the UP. "Shou'd it be said, that we have experience, that the same power continues united with the same object, and that like objects are endow'd with like powers, I wou'd renew my question, *why from this experience we form any conclusion beyond those past instances, of which we have had experience*" (T 1.3.6.10). Now Hume's previous argument (T 1.3.6.6) about the UP will recur again after all. There is no noncircular argument for the UP, and thus no noncircular argument for the claim that similar objects always possess similar powers. The invocation of "powers" does not help escape the problem that reason does not produce the supposition of the UP.

In the last paragraphs of the section, Hume explains the conclusion that causal inferences are produced by the imagination (T 1.3.6.12–16). In his words, "When the mind, therefore, passes from the idea or impression of one object to the idea or belief of another, it is not determin'd by reason, but by certain principles, which associate together the ideas of these objects, and unite them in the imagination" (T 1.3.6.12). Three non-rational associative principles link perceptions together in the imagination: resemblance, contiguity, and causation. Two objects stand in the philosophical relation of causation when they are contiguous, when the action of one precedes the action of the other, and when they are constantly conjoined in our past experience. Yet the philosophical relation of causation affords no grounds for a rational inference to unobserved effects. Causal inference only occurs because the ideas of the two objects are causally connected in the imagination, such that the impression of one produces belief in the other—for no rational reason. The natural association of ideas in our minds affords causal inferences, not the philosophical relation of causation between objects.

CAUSAL INFERENCE IS DEFEASIBLY JUSTIFYING

The broad outline of Hume's argument in *Treatise* 1.3.6 is fairly clear, as is his conclusion that causal inference is not produced by reason but by associative principles of the imagination. However, commentators continue to debate whether that conclusion has any normative epistemic significance, and if so, what it is. There are three basic positions on the epistemic status of the conclusions of causal inferences by the end of *Treatise* 1.3.6.[8] Hume either regards them defeasibly justified, or as not defeasibly justified, or else he has not yet registered an epistemic assessment of them at all.

8. These three positions could each receive further nuance in terms of kinds of causal inference, degrees of evidence, and so forth.

According to many scholars, by the end of *Treatise* 1.3.6, Hume has shown that the conclusions of causal inferences are not defeasibly justified.[9] For Hume, they say, a belief produced merely by associative principles of the imagination, unsupported by good arguments from observation or memory, is not epistemologically justified. Production by reason is a necessary as well as sufficient condition for justification.

On the other hand, a large number of recent interpreters deny that Hume intends this flatly skeptical conclusion at *Treatise* 1.3.6.[10] Interestingly, Hume's early Scottish critics also seem not to have construed *Treatise* 1.3.6 as a skeptical argument.[11] After all, Hume keeps right on making inductive inferences in his work. He surely endorses the "Rules by which to judge of cause and effects" (T 1.3.15) as well as the rules of probable inference he expounds in *Treatise* 1.3.11–13. It appears that even at the end of *Treatise* 1.3.6, Hume holds that the conclusions of causal inferences are justified.[12]

According to some non-skeptical interpreters, Hume can continue to maintain that causal inference is justifying after *Treatise* 1.3.6 only because he changes his usage of the term "reason." Louis E. Loeb, for example, points out that even at the end of *Treatise* 1.3.6, he writes of "our reasonings from that relation" (T 1.3.6.15) and says that "we are able to reason upon" the natural relation of causation (T 1.3.6.16).[13] How can these statements be rec-

9. Stroud, *Hume*, 52–55; Kenneth Winkler, "Hume's Inductive Skepticism," in *The empiricists: critical essays on Locke, Berkeley, and Hume*, edited by Margaret Atherton (Lanham, MD: Rowman & Littlefield, 1999), 183–212; Fogelin, *Hume's Skepticism*, chapter 4 and Appendix A, as well as "Hume's Skepticism," in *The Cambridge Companion to Hume*, 2nd ed., edited by David Fate Norton and Jacqueline Taylor (Cambridge: Cambridge University Press, 2009), 214–21; De Pierris, *Ideas, Evidence, and Method*, chapter 6, 197–258.

10. Loeb cites many commentators who "note Hume's favorable attitude toward causal inference" at least in Part 3 of Book 1 of the *Treatise*. Loeb, *Stability*, 38n1. Loeb correctly writes: "The evidence of Hume's approval of causal inference at these stages of Book 1 is overwhelming . . . There is no doubt that Hume discovered the materials used to formulate the problem of induction familiar to twentieth-century philosophy. Hume did not, however, intend the argument of I.iii.6, as deployed in the *Treatise*, to have any skeptical weight or force." Ibid., 38.

11. In his chapter "Of Probable Reasoning," Thomas Reid gives basically same account of probable reasoning as Hume himself does, arguing that our instinctual trust in the uniformity of nature is epistemologically justified even though unsupportable by argument. *Essays on the Intellectual Powers of Man*, in *The Works of Thomas Reid*, vol. 1, Essay 6, chapter 7, 481–84. "General rules may have exceptions or limitations which no man ever had occasion to observe. . . . But we are led by our constitution to rely upon their continuance with as little doubt as if it was demonstrable." Ibid. 484b. In the very next chapter, "Of Mr. Hume's Scepticism with Regard to Reason," Reid subjects to criticism only Hume's argument from *Treatise* 1.4.1. Ibid., 484–90. Reid never passes up an opportunity to counter Humean skepticism, perceived or real. If he had read *Treatise* 1.3.6 as a skeptical argument, he almost certainly would taken up the cudgels against it. Loeb points out that even James Beattie, author of *An Essay on the Nature and Immutability of Truth*, one of Hume's least charitable Scottish commonsense critics, never construes the argument of *Treatise* 1.3.6 as a skeptical threat. Loeb, *Stability*, 50–51.

12. See especially Janet Broughton's seminal article "Hume's Skepticism about Causal Inferences," *Pacific Philosophical Quarterly* 64 (1983), 3–18. Cf. also Noonan, *Hume on Knowledge*, 121–31; Loeb, *Stability*, 43–47; Schmitt, *Hume's Epistemology*, 131–242.

13. Loeb, *Stability*, 53–54.

onciled with Hume's denial that reason produces causal inferences? Hume must change the meaning with which he uses the term "reason." Whereas previously to the main argument of *Treatise* 1.3.6, Hume uses "reason" to refer only to a non-associative faculty, afterward he uses "reason" to include the associative mechanism whereby we make causal inferences. Hume has a pretheoretical commitment to the positive epistemic status of causal inference, and "reason" conveys epistemic endorsement.[14] *Treatise* 1.3.6 shows that *if* we define "reason" as a non-associative faculty, then causal inference is unreasonable. But (of course) causal inference is reasonable. Therefore, "reason" is an associative faculty.

Peter Millican argues similarly that Hume changes his definition of "reason" in order to maintain his view that induction is produced by justifying cognitive faculties.[15] Millican writes that "induction is such a central cognitive process that it ought *by definition* to be an operation of *reason*," and hence "*custom*—as the underlying process that drives induction—should itself be part of *reason*."[16] This leads Hume to reassign PIU principles (including custom) to "reason." But it also creates a terminological muddle, since he also continues to insist that inductive inference "is not determin'd by reason."

Both Loeb and Millican labor under the false assumption that "reason" is Hume's key term of epistemic approbation, and that production by reason is necessary for defeasible justification. If production by reason is necessary to justification, and if Hume continues to regard induction as defeasibly justifying, then he must, after all, regard reason as the cause of induction. But as I argued in chapter 1, reason is not the only PIU principle; it is not the only defeasibly justifying faculty. As I argued in chapter 2, production by reason is sufficient but not necessary for justification. The instinct that causes inductive inferences need not be part of "reason" in order to be a defeasibly justifying, PIU principle. So the main motive for seeing a change in Hume's terminology after *Treatise* 1.3.6 vanishes.

Furthermore, there is no textual evidence for a shift in meaning of "reason."[17] Garrett rightly argues that "Hume uses the term 'reason' quite univocally to refer to the inferential faculty—a faculty that produces two kinds

14. "Such terms as 'reason' and the 'understanding' convey favorable epistemic assessments, ones Hume thinks appropriate to apply to causal inference . . . the entire rationale for the shifts in terminology is to retain the favorable connotations of cognates of 'reason' in application to causal inference." Loeb, *Stability*, 54.

15. Millican, "Hume's 'Scepticism,'" 87.

16. Ibid.

17. Garrett makes this point while criticizing earlier anti-deductivist readings that also stipulated a special meaning of "reason" in *Treatise* 1.3.6. Garrett, *Cognition*, 84–85. Millican notes that "Garrett has consistently urged this last point against rival interpretations: that Hume's famous argument gives little internal clue that he is employing some special notion of reason which he aims to reject . . . the lack of any obvious and deliberate ambiguity or equivocation on Hume's part has remained by far the strongest weapon in Garrett's armoury." Millican, "Hume's 'Scepticism,'" 74, 75.

of arguments, demonstrative and probable."[18] It is not necessary to attribute a change in the meaning of "reason" to explain Hume's consistent assertions that we engage in causal *reasoning*. As Garrett argues, Hume never denies that the transition from observational impression to belief in an unobserved idea is an act of reasoning. His famous argument is meant to show that that act of reasoning is not itself produced by another antecedent act of reasoning. Causal inferences are bits of reasoning which are produced by the supposition of the UP, and that supposition is produced by custom, not by reasoning or argument.[19]

Garrett and David Owen are the chief representatives of a third view that sees Hume, at the end of *Treatise* 1.3.6, as neutral or uncommitted regarding the justificatory status of the conclusions of causal inferences. Garrett and Owen deny that Hume registers any epistemic evaluations at all until he utters the Title Principle at *Treatise* 1.4.7.11.[20] Hume neither affirms nor denies the positive epistemic status of causal inference at 1.3.6 or in the rest of Part 3.[21] As Garrett explains, "reason," for Hume, is not a normative epistemic term, but "simply the name that Hume, as cognitive psychologist, consistently employs for the general faculty of making inferences or producing arguments."[22] In *Treatise* 1.3.6 Hume merely observes one role that reason does not play: it does not determine the mind to draw causal inferences. Owen adds that Humean reason is a faculty with norms of proper and improper use which are spelled out in places like *Treatise* 1.3.15.[23] The norms of the proper use of reason qua faculty should not be identified with epistemic norms per se. We must "distinguish reason or the understanding from the broader philosophy in which it is embedded," which tells us when and how the faculty should be used at all.[24] For both Garrett and Owen, the broader epistemology in which reason is embedded is expressed in the Title Principle. On their reading, the Title Principle gives a qualified endorsement to the deliverances of the faculty of reason: they are epistemologically justified if and only if we have a propensity to assent to them.

I agree with Garrett and Owen that Hume's claims about reason and its products are descriptive claims about a cognitive mechanism without intrinsic epistemic significance. "Reason" is not an epistemic success term for Hume. Both skeptical and anti-skeptical interpreters have wrongly assumed that production by reason is a necessary condition of justification. Skeptical interpreters have

18. Garrett, *Cognition*, 94.
19. Garrett, *Hume*, 178–79.
20. Garrett, *Cognition*, 232–37; Owen, *Hume's Reason*, chapter 9.
21. Garrett, *Cognition*, chapter 4, especially 94–95; Owen, *Hume's Reason*, chapter 6.
22. Garrett, *Cognition*, 92.
23. Owen, *Hume's Reason*, 206.
24. Ibid. 221

assumed that since causal inferences are not produced by reason, their conclusions are not justified. Loeb and other anti-skeptical interpreters have assumed that since causal inferences do have justified conclusions, then those inferences must be produced by reason after all, despite the conclusion of *Treatise* 1.3.6. But contrary to both these approaches, we cannot assume a priori that production by reason is either necessary or sufficient for justification.

On the other hand, the purely descriptive reading of *Treatise* 1.3.6 ignores Hume's true epistemic success terms, the terminology of "philosophy." From the beginning of the *Treatise* all the way through Part 3 of Book 1, there is no question that the philosophers approve of inductive inference and regard its products as defeasibly justified. "The customary transition from causes to effects, and from effects to causes" is Hume's preeminent example of the principles of the imagination "which are permanent, irresistible, and universal," and of which the philosophers approve (T 1.4.4.1–2). Just probable belief-forming processes are described as those which "are receiv'd by philosophers, and allow'd to be reasonable foundations of belief and opinion," and are explicitly contrasted with "unphilosophical probability" (T 1.3.13.1). The rules for causal judgment, which norm both natural and moral philosophy (T 1.3.15.11), are constitutive of philosophy in the general sense.[25]

Contrary to the view of Garrett and Owen, Hume does not defer epistemic questions until *Treatise* 1.4.7. Both before and after *Treatise* 1.3.6, Hume relies unreservedly on causal inference—probable reasoning—for the development of his science of man. Hume (with the philosophers) approves of causal inference prior to developing any view about what cognitive mechanisms produce it. Hume (with the philosophers) continues to approve of causal inference after discovering that it is produced by custom, not reason. He regards his analysis of causal inference as a groundbreaking discovery. But it is a discovery about the descriptive psychology of probable reasoning, not about its epistemic status, which was taken for granted all along.

THE DEFEAT OF REASON'S PRODUCTS IN *TREATISE* 1.4.1

Although Hume thinks causal inference produces defeasibly justified beliefs, inasmuch as they are produced by permanent, irresistible, and universal propensities of the imagination, he also admits that these beliefs face epistemic defeat from reason. Hume's defeater argument against inductive conclusions occurs in *Treatise* 1.4.1, "Of scepticism with regard to reason." My main ob-

25. Hume mentions that the fourth rule, that "The same cause always produces the same effect, and the same effect never arises but from the same cause," is "the source of most of our philosophical reasonings" (T 1.3.15.6).

jective in this section is to show that, contrary to non-skeptical interpreters, Hume views these arguments as successful skeptical challenges to the epistemic justification of core beliefs. A secondary goal is to show that Hume's skeptical argument in *Treatise* 1.4.1 is a defeater argument from reason. The argument presupposes that reason is defeasibly justifying, even if its products are all ultimately defeated.

The skeptical argument in *Treatise* 1.4.1 provides an undermining defeater for the conclusion of any piece of demonstrative reasoning. Along the way, the argument also provides an undermining defeater for any inductive inference. All told, the argument undermines the epistemic justification of all the products of the faculty of reason—that is, for the conclusions of all inferences, demonstrative or probable. Recall that an undermining defeater for some belief *P* is not a reason to believe not-*P*, but rather a reason not to believe that *P*. Hume intends his skeptical argument against reason to show that we epistemologically ought to iteratively decrease our confidence in any of reason's conclusions until we no longer believe them.

The skeptical argument against demonstrative conclusions proceeds in two stages. First, Hume argues that although the rules of "demonstrative sciences" are infallible, humans only fallibly follow them in specific acts of reasoning (T 1.4.1.1–3). Our past record of errors obliges us to admit that, for any given demonstrative inference we make, at best we are only very likely to have drawn a correct conclusion. In light of the fallibility of our demonstrative reasoning, Hume pronounces the following epistemic norm: "We must, therefore, in every reasoning form a new judgment, as a check or control on our first judgment or belief" (T 1.4.1.1). Proportionate to our past track record as fallible reasoners ("a kind of history of all the instances, wherein our understanding has deceiv'd us, compar'd with those, wherein its testimony was true," ibid.), we should adjust our confidence level in the conclusion of a demonstration downward from certainty to a probability of less than one. Hume says that in this way, all "knowledge" degenerates or resolves itself into probability (T 1.4.1.3). By this he means that even those beliefs, arising from intuition or demonstration (T 1.3.1.2), of which we are initially certain, become merely probable when we duly reflect on them.

In the second stage of the argument, Hume says that our probable reasoning is also fallible. He therefore pronounces another epistemic norm (or rather, makes a fresh application of the one he already stated): "In every judgment, which we can form concerning probability, as well as concerning knowledge we ought always to correct the first judgment, deriv'd from the nature of the object, by another judgment, deriv'd from the nature of the understanding" (T 1.4.1.5). Hume indicates that, proportionate to our past successes and failures in probable reasoning, upon reflection we ought to adjust our confidence in our probable conclusions downward to some degree. If we are only 99 percent confident that we should be 99 percent confident of some proposition *P*, then by Hume's lights we ought only to be (say) 98 percent confident of *P*. Our higher-order doubts about our ability to make good first-order probable judgments should decrease our first-order confidence.

Hume does not make it entirely clear why he thinks that higher-order judgments about our reasoning abilities should lower the probability of our first-order judgments. Perhaps the most common criticism of Hume's argument in *Treatise* 1.4.1 is that at precisely this point, he confuses the contents of probability beliefs with the strength of those beliefs; that is to say, he conflates objective and subjective probability.[26] As Fogelin puts it, "However certain or uncertain we are about our ability to calculate probabilities, if a proposition has a certain probability, that (tautologically) is the probability it has."[27]

Several authors suggest that Hume is not thinking in terms of objective probability at all, but only in terms of subjective confidence levels.[28] On this reading, the second-order realization that there is some slight chance that our first-order confidence level is wrong, should lead us to lower our first-order confidence level somewhat. But again the question may be asked, "Why should the realization of our fallibility lead us to lower our first-order confidence levels rather than raising them? If we have erred, our confidence may just as well be too low as too high."[29]

Michael P. Lynch proposes that a second-order realization of first-order fallibility could reasonably lead us to widen our confidence interval on both sides.[30] For example, if my original first-order confidence in *P* was 0.8, then when I recall my past errors, I should widen my confidence interval to somewhere between 0.7 and 0.9. Admittedly, Lynch's reconstruction differs from Hume's text. Hume says only that we should lower our confidence levels, not that we should widen our confidence intervals. But Lynch might be stating precisely what Hume states inchoately. For my purposes, it does not matter whether Lynch's interpretation is the best one or not. For whatever reason or

26. Frequently cited on this point are Ian Hacking, "Hume's Species of Probability," *Philosophical Studies*, 33(1), 1978, 30, and Robert Imlay, "Hume's 'Of scepticism with regard to reason': A Study in Contrasting Themes," *Hume Studies* 7(2), 1981, 124–25.

27. Fogelin, *Hume's Skepticism*, 18. Fogelin credits Thomas Reid with having made this criticism even earlier. Fogelin, *Hume's Skepticism*, 174, citing Reid, *Essays on the Intellectual Powers of Man*, (Cambridge, MA: MIT Press, 1969), 746. See also Loeb, *Stability*, 223–24; Schmitt, *Hume's Epistemology*, 323n9.

28. See William Edward Morris, "Hume's Scepticism About Reason," *Hume Studies* 15(1), April 1989, 40–53; Lynch, "Hume and the Limits of Reason," 90–96; Owen, "Scepticism with Regard to Reason," in *The Cambridge Companion to Hume*, ed. Donald C. Ainsle and Annemarie Butler (New York: Cambridge University Press, 2015), 109–117.

29. Lynch, "Hume and the Limits of Reason," 93, citing Mikael M. Karlsson, "Epistemic Leaks and Epistemic Meltdowns: A Response to William Morris on Scepticism with Regard to Reason," *Hume Studies* 16.2 (1990), 126.

30. Lynch, "Hume and the Limits of Reason," 94.

for none at all, Hume thinks that second-order assessments of the correctness of our first-order probable reasoning abilities should decrease the confidence with which we hold our first-order probable beliefs.

Granting Hume this key claim—that I ought to lower my confidence in *P* in proportion to how likely I think it is that I wrongly assessed the probability that *P*—clears the way for his destructive iterative skeptical argument. For any first-order probable belief *P*, second-order reflection on my fallibility should lead me to lower my confidence in *P*. Third-order reflection on the fallibility of my second-order judgment *Q* should lead me to lower my confidence in *Q* and thus also in *P*. This iterative process of evaluating the reliability of our successive corrective judgments, and decreasing our confidence in our original judgment, goes on indefinitely (T 1.4.1.4–5). If we continue this iterative process of corrective judgments, our confidence in the original demonstrative judgment at last diminishes below the threshold of belief (T 1.4.1.6). An infinite number of higher-order reflections of this sort should diminish confidence in *P* until I have no confidence in *P*.

The bottom line is that if we follow our epistemic norms, which require us to decrease our confidence in our inferences based on our fallibility as thinkers, then we will eventually suspend all of our reason-based judgments. The first part of the argument is supposed to show that all "knowledge" (certainty based on demonstration) degenerates into probability. Iterative higher-order judgments about the reliability of our demonstrative reasoning should lead us to hold its conclusions as merely probable. The second part of the argument is supposed to show that if we reflect as we ought to on our probable beliefs, we suspend them all.

My aim in examining this argument is to determine Hume's view of it, not to assess its soundness. Although the argument does have its defenders, it is widely regarded as fallacious.[31] Sound or not, it shows us the sources and nature of Hume's skepticism.

Hume's argument presupposes that our inferences, probable and demonstrative, are defeasibly justifying. The corrective judgments which eventually undermine our confidence in all of our conclusions are themselves produced by inductive inferences regarding the reliability of our faculties. These inductive inferences are based upon "a kind of history of all the instances, wherein our understanding has deceiv'd us, compar'd with those, wherein its testimony was just and true" (T 1.4.1.1). Based on past experience, Hume demands that we form a judgment about the likelihood that our faculty of reason has produced an accurate result in the present case. Hume takes it for granted

---

31. As noted above, those who emphasize that the argument is deeply flawed include Fogelin, Loeb, and Schmitt. Fogelin, *Hume's Skepticism*, 13–24; Loeb, *Stability*, 223–29; Schmitt, *Hume's Epistemology*, 323. Morris, Lynch, and Owen attempt to rehabilitate the argument to some degree. Morris, "Hume's Scepticism About Reason," 40–53; Lynch, "Hume and the Limits of Reason," 90–96; Owen, "Scepticism with Regard to Reason," 109–117.

that inferences from past experience are perfectly respectable, epistemologically speaking. It is just because such inferences are justifying that iterated corrective judgments defeat all reason's products.

OBJECTIONS TO THE SKEPTICAL INTERPRETATION OF *TREATISE* 1.4.1

Although it might seem natural to treat "Of scepticism with regard to reason" as a section in which Hume argues for epistemic skepticism, many commentators read it in a non-skeptical fashion.[32] They make at least four important points in favor of their reading.

First, non-skeptical interpreters contend that Hume himself rejects the epistemic norm which requires us to successively correct our inferential confidence levels.[33] Hume expresses this norm (which I will call the Corrective Norm) in statements I quoted already above:

> We must, therefore, in every reasoning form a new judgment, as a check or controul on our first judgment or belief. (T 1.4.1.1)

> In every judgment, which we can form concerning probability, as well as concerning knowledge, we ought always to correct the first judgment, deriv'd from the nature of the object, by another judgment, deriv'd from the nature of the understanding. (T 1.4.1.5)

Non-skeptical readers argue that this norm should not lead to skepticism because Hume does not endorse it, at least not without qualification. For example, Garrett asserts that Hume's own epistemic norm, the Title Principle, limits the extent to which we ought to follow the Corrective Norm.[34] Once

32. For epistemologically non-skeptical readings of *Treatise* 1.4.1, see Morris, "Hume's Scepticism About Reason," and "Hume's Conclusion"; Lynch, "Hume and the Limits of Reason"; Garrett, *Cognition*, 229–32, and *Hume*, 223–26; Mounce, *Hume's Naturalism*, 49–52; Schmitt, *Hume's Epistemology*, 317–40; Owen, "Scepticism with Regard to Reason." For epistemologically skeptical readings, see Fogelin, *Hume's Skepticism*, 13–24; Noonan, *Hume on Knowledge*, 12, 161–62; Loeb, *Stability*, 85–86, 223–29; Meeker, "Hume: Radical Skeptic" and especially "Hume's Iterative Probability Argument: A Pernicious *Reductio*," in *Journal of the History of Philosophy* 38.2 (April 2000), 221–38; Broughton, "Hume's Naturalism and His Skepticism," 431.

33. Morris says that Hume rejects this model of "rationally reflective epistemic agents." Morris, "Hume's Scepticism about Reason," 56–58, and "Hume's Conclusion," 102–6. Baier's view is very close to Morris's (it treats the argument as a reductio ad absurdum) but is less carefully worked out. Baier, *Progress*, 287; cf. 184. Mounce holds that the skeptical argument against reason in *Treatise* 1.4.1 arises only for an "autonomous" use of reason which Hume himself rejects. Mounce, *Hume's Naturalism*, 49–52. Schmitt says that Hume does not endorse this "Norm of Reduction." Schmitt, *Hume's Epistemology*, 323–24.

34. Garrett, *Cognition*, 235. Just as a reminder, the Title Principle says "Where reason is lively, and mixes itself with some propensity, it ought to be assented to. Where it does not, it can never have any title to operate upon us" (T 1.4.7.11).

our self-corrective reasoning ceases to be lively or mix with a propensity, the Title Principle permits us to ignore it. Self-corrective reasoning ceases to be lively long before it diminishes our inferential confidence levels below the belief threshold (as Hume explains in T 1.4.1.10–11). The Corrective Norm requires us to adjust our confidence levels in "every" new inferential judgment we form. The Title Principle permits us to retain our confidence. When the two norms conflict, Hume favors the Title Principle.

I agree that if the Title Principle is Hume's final epistemic norm, overriding the Corrective Norm, then he might plausibly take it to be an epistemic solution to skepticism about inferences. However, the Title Principle appears only much later (T 1.4.7.11). The non-skeptical reading of *Treatise* 1.4.1 that hinges upon an epistemic construal of the Title Principle can only be fully addressed in the context of a full interpretation of *Treatise* 1.4.7. In *Treatise* 1.4.1 Hume endorses only the Corrective Norm without any hint of restriction. Furthermore, I argue later that the Title Principle is not an epistemic norm (a norm of philosophy), but a merely practical norm. The Title Principle practically permits us to ignore the Corrective Norm under certain conditions, but it does not epistemologically permit us to do so. It eliminates the threat of practical skepticism, but not of epistemic skepticism.

Second, non-skeptical readers point out that Hume specifically denies that either he or anyone else can suspend his own beliefs that are produced by reasoning.

> Shou'd it here be ask'd me, whether I sincerely assent to this argument, which I seem to take such pains to inculcate, and whether I be really one of those sceptics, who hold that all is uncertain, and that our judgment is not in *any* thing possest of *any* measures of truth and falsehood; I shou'd reply, that this question is entirely superfluous, and that neither I, nor any other person was ever sincerely and constantly of that opinion. (T 1.4.1.7)

But if Hume does not suspend his beliefs, he must not think that his beliefs are epistemically unjustified. The fact that Hume does not suspend his reason-based beliefs shows that in general he does not think the argument is sound. More specifically, he does not regard himself as obliged to follow the Corrective Norm and reduce his confidence levels below the belief threshold.[35]

Hume certainly does insist that no one is psychologically capable of suspending belief in the conclusions of reasoning. But it does not follow that he does not think we *ought* to suspend our reason-based beliefs. We may simply be psychologically incapable of doing our epistemic duty.[36]

35. Morris, "Hume's Scepticism about Reason," 55; Schmitt, *Hume's Epistemology*, 323–24.

36. Schmitt thinks that, because Hume does not suspend judgment in inferential beliefs, he does not endorse the Norm of Reduction (which I have called the Corrective Norm). Schmitt, *Hume's Epistemology*, 323–24. But Schmitt himself points out a major textual problem with his view. Hume says that the reason he does not suspend his beliefs is not because he disbelieves the Norm of Reduction,

Hume distinguishes the psychological strength of a belief from its epistemic justification. Beliefs produced by "education" are frequently ineradicable, but they are nevertheless condemned by philosophers, since they often contradict reason and themselves (T 1.3.9.17–19). The vulgar frequently form prejudicial beliefs of which, whatever their psychological strength, the philosophers disapprove (T 1.3.13.7–18). With respect to *Treatise* 1.4.1, we must distinguish between two questions. The first question is, "Psychologically, does the iterative application of the Corrective Norm cause us to suspend our reason-based beliefs?" Hume answers this question in the negative. But this does not entail any particular answer to the second question, which is, "Epistemologically, should the iterative application of the Corrective Norm cause us to suspend our reason-based beliefs?" Hume answers this second question affirmatively: "all the rules of logic require a continual diminution, and at last a total extinction of belief and evidence" (T 1.4.1.6).

Third, non-skeptical readers claim that Hume's primary aim is to make a point in descriptive psychology; whatever evaluative epistemic implications his conclusions may have (if any) are secondary.[37] Hume himself seems to say as much:

> My intention then in displaying so carefully the arguments of that fantastic sect, is only to make the reader sensible of the truth of my hypothesis, *that all our reasonings concerning causes and effects are deriv'd from nothing but custom; and that belief is more properly an act of the sensitive, than of the cogitative part of our natures* . . . If belief, therefore, were a simple act of the thought, without any peculiar manner of conception, or the addition of a force and vivacity, it must infallibly destroy itself, and in every case terminate in a total suspense of judgment. (T 1.4.1.8)

In the preceding quotation Hume announces that he intends "only" to prove two psychological theses. First, he intends to prove that causal inferences are derived from nothing but custom. Second, he intends to prove that the difference between entertaining and believing a proposition lies only in the degree of force and vivacity. If these theses were not true, then we would recursively self-correct our judgments until we had no reason-based beliefs. If causal

but because he is psychologically unable to sustain his confidence in a long chain of reasoning. Ibid. 334–35. Schmitt himself says "I see no way to defuse the apparent inconsistency here"—that is, the inconsistency between Hume's explanation of the issue, and Schmitt's own. Ibid. 335.

37. Garrett, *Cognition*, 227–28; Owen, "Scepticism with Regard to Reason," especially 118–21; Schmitt, *Hume's Epistemology*, 317–18. Morris says that Hume's goal is to reject a certain "intellectualist" model of the mind. Morris, "Hume's Scepticism about Reason," 55–58. This sounds like a project in descriptive psychology. But Morris's Hume really rejects a certain model of "rationally reflective epistemic agents," which is not merely descriptive of how reason does work, but how it ought to work. This is especially clear in Morris, "Hume's Conclusion," 102–6. Morris's Hume does not merely reject a set of descriptive claims, but rejects the Corrective Norm.

inference were derived from something more or other than associative principles, and if assent amounted to something other than the transference of force and vivacity in accordance with those associative principles, then the length and abstruseness of a chain of reasoning would constitute no barrier to assent. But in fact, the length and abstruseness of a chain of reasoning do constitute a barrier to assent, and we do retain our reason-based beliefs. So Hume's two psychological theses about causal reasoning and belief are confirmed.

I gladly agree that Hume does intend to defend these two theses of descriptive psychology. But this concession does not preclude that Hume is also making a skeptical epistemic point. Hume means his argument to show that we psychologically cannot follow a skeptical chain of reasoning that we epistemologically ought to follow. Our inability to follow the Corrective Norm (which Hume endorses) proves that belief is a peculiar manner of conceiving ideas, with sufficient force and vivacity. Force and vivacity cannot transfer through long chains of argument (howsoever sound these arguments may be) (T 1.4.1.9–10). So we cannot hold all of the beliefs that we ought to. Hume says:

> But as experience will sufficiently convince any one, who thinks it worth while to try, that tho' he can find no error in the foregoing arguments, yet he still continues to believe, and think, and reason as usual, he may safely conclude, that his reasoning and belief is some sensation or peculiar manner of conception, which 'tis impossible for mere ideas and reflections to destroy. (T 1.4.1.8)

There is "no error in the foregoing arguments": they are sound, and epistemologically we ought to accept them. But we are psychologically unable to submit to their conclusion, and this confirms Hume's theory of causal reasoning and belief.[38]

Fourth, non-skeptical readers insist that in *Treatise* 1.4.1, Hume uses the word "evidence" as a synonym for "evidentness," not as a normative epistemic term.[39] Put another way, these readers take "evidence" to denote the psychological strength of a belief, not its epistemologically justifying grounds. This crucially determines what Hume means when he says that the recursive self-corrections of reason diminish the "original evidence" of the first inference (T 1.4.1.8, 1.4.1.9). He says that

---

38. This is basically Meeker's reading. Meeker, "Hume's Iterative Probability Argument," 233–38. Meeker states the point in terms of "evidence," which, as I discuss below, I take in a somewhat different sense than he does.

39. Garrett, *Cognition*, 228; Owen, *Hume's Reason*, 185–88; Owen, "Scepticism with Regard to Reason," 102, 116, 131–32n33. Meeker argues for an epistemic meaning of "evidence." Meeker, "Hume's Iterative Probability Argument," 224–27.

> When I reflect on the natural fallibility of my judgment, I have less confidence in my opinions, than when I only consider the objects concerning which I reason; and when I proceed still farther, to turn the scrutiny against every successive estimation I make of my faculties, all the rules of logic require a continual diminution, and at last a total extinction of belief and *evidence*. (T 1.4.1.6, italics mine)

Hume uses "evidence" here interchangeably with "confidence in my opinions," and with the diminution of "belief." The argument proves therefore that the recursive self-corrections of reason extinguish the strength of our reason-based beliefs, but not their justification.

The debate over the meaning of "evidence" is not as important as it may seem. I think that Kevin Meeker rightly points out that Hume uses "evidence" as a term of epistemic evaluation at least sometimes, as in his famous assertion that "A wise man, therefore, proportions his belief to the evidence" (EHU 10.4).[40] But it does not follow that Hume always uses "evidence" in this way. In *Treatise* 1.4.1, I incline toward the view of Garrett and Owen that "evidence" means only belief-strength or confidence.

However, as with preceding two objections, the mere fact that Hume is making a psychological point does not preclude his also making a distinct epistemic point. In the passage quoted above, the epistemological point stands, even granting that "evidence" means only "belief-strength." In the statement "all the rules of logic require a continual diminution, and at last a total extinction of belief and evidence," the normative epistemic force lies in the expression "all the rules of logic require," not in the word "evidence." The statement means "Epistemic norms require a total extinction of confidence."

Even though all of reason's products meet with defeat, it is still a PIU principle. Hume is at pains to point out that even in the face of his skeptical argument, reason cannot be suppressed.

> Nature, by an absolute and uncontroulable necessity has determin'd us to judge as well as to breathe and feel; nor can we any more forbear viewing certain objects in a stronger and fuller light, upon account of their customary connexion with a present impression, than we can hinder ourselves from thinking as long as we are awake, or seeing the surrounding bodies, when we turn our eyes towards them in broad sunshine. (T 1.4.1.7)

Reason is "a faculty, which nature has antecedently implanted in the mind, and render'd unavoidable" (T 1.4.1.7). Regardless of the defeat of its deliverances, reason (demonstrative and probable) is a permanent, irresistible, and universal principle of the mind. It therefore receives the sanction of the

40. Meeker, "Hume's Iterative Probability Argument," 224.

philosophers in *Treatise* 1.4.4.1 (a passage, we should note, that follows the destructive skeptical argument of T 1.4.1). Its deliverances are defeasibly justified, even though they are all ultimately unjustified.

In this chapter I have argued that Hume does not intend a skeptical conclusion at *Treatise* 1.3.6. He regards causal inference as defeasibly justifying even though it is produced by custom rather than reason. But Hume does intend a skeptical conclusion at *Treatise* 1.4.1. All conclusions of demonstrative and probable reasoning face insuperable undermining defeaters from reason itself. Since reason is a PIU principle, the products of reason are defeasibly justified, but consequently defeated. However, we are psychologically incapable of suspending our belief in reason-based conclusions. This psychological incapacity also happens to confirm Hume's descriptive account of the nature of belief.

Chapter Four

The Senses and Skepticism in the *Treatise*

In the last chapter I argued that beliefs in the conclusions of reasoning (probable and demonstrative) are defeasibly justified but succumb to undermining defeaters set forth in *Treatise* 1.4.1. In this chapter I show that Hume treats belief in external objects the same way. I argue that the vulgar belief in continued and distinct existences (bodies), as Hume describes it in *Treatise* 1.4.2, "Of scepticism with regard to the senses," is defeasibly justified. Prior to and apart from the rebutting defeater that Hume brings forward as an argument from perceptual relativity in paragraphs 44 and 45, the vulgar belief is perfectly in order, philosophically speaking.[1] But that rebutting defeater creates an insuperable skeptical problem for all belief in body. The other three other versions of belief in body that Hume discusses (the double existence theory, the ancient view of substances, and the modern theory of primary and secondary qualities) lack even defeasible justification. Furthermore, they all face defeaters from reason. So no version of belief in body is philosophically justified overall.

I begin the chapter with an overview of the complex textual terrain of *Treatise* 1.4.2–1.4.4. Then I take up the justificatory status of the four versions of belief in body that Hume discusses.

1. Similar views are defended by Fred Wilson, *The External World and our Knowledge of It: Hume's Critical Realism, an Exposition and a Defence* (Toronto: University of Toronto Press, 2008), 597–98, and Ainslie, *Hume's True Scepticism*, 106–8. Garrett does not explicitly comment on the defeasible justification of the vulgar belief, but he seems to grant it. Garrett, *Hume*, 97–101. On Garrett's view, the vulgar belief does not involve a logical contradiction. It involves a mistake which is exposed by experimental evidence. None of these scholars explicitly distinguish between defeasible and overall justification.

AN OVERVIEW OF *TREATISE* 1.4.2–1.4.4

The discussion of belief in body begins in *Treatise* 1.4.2, "Of scepticism with regard to the senses." Belief in body is belief in continued and distinct existences (T 1.4.2.2). A continued existence is one that exists even when it is not present to our senses. A distinct existence is spatially external to the mind and does not depend on the mind causally for its existence or operations.

Hume examines four different versions of belief in body. (1) First he takes up the *vulgar* belief in body (T 1.4.2.14–45; cf. EHU 12.7–9). The vulgar "attribute a distinct continu'd existence to the very things they feel or see" (T 1.4.2.14). The vulgar assume, that is, that the very perceptions that perceivers have in an act of perception continue to exist when no perceiver is having them. (2) Philosophers who reject the vulgar view adopt the *system of double existences*, according to which our perceptions, which are neither continued nor distinct, are caused by objects that exactly resemble the perceptions they cause (T 1.4.2.46–55). These exactly resembling objects are continued and distinct existences. (3) The "*antient philosophy*" construes bodies as substances with substantial forms, accidents, and occult qualities (T 1.4.3). (4) The "*modern philosophy*" is the Lockean view of primary and secondary qualities (T 1.4.4; cf. EHU 12.10–15). On this account, "colours, sounds, tastes, smells, heat and cold" are "nothing but impressions in the mind, deriv'd from the operation of external objects, and without any resemblance to the qualities of the objects" (T 1.4.4.3). Perceptions of primary qualities are supposedly effects of resembling qualities in objects themselves, objects which are continued and distinct existences.

The title of *Treatise* 1.4.2 ("Of *scepticism* with regard to the senses," emphasis mine) indicates that Hume is arguing that belief in body is unjustified. As usual, however, he determines the justificatory status of the belief by investigating its source—the cognitive mechanisms that produce it: "The subject, then, of our present enquiry is concerning the *causes* which induce us to believe in the existence of body" (T 1.4.2.2). Regardless of what we conclude about the causes or justification of belief in body, we are compelled psychologically to hold this belief, in one form or another: "Nature has not left this to his [the sceptic's] choice" (T 1.4.2.1). Since we cannot help but believe in bodies, "'tis in vain to ask, *Whether there be body or not?* That is a point, which we must take for granted in all our reasonings" (T 1.4.2.1).

THE VULGAR BELIEF IN BODY

The Defeasible Justification of the Vulgar Belief in Body

An influential tradition reads Hume as denying the defeasible justification of the vulgar belief in body.[2] Not everyone in this tradition explicitly distinguishes between defeasible and overall justification. But they all see the vulgar belief as condemned by and from its very origins in the human mind, not simply by subsequent evidence. The cognitive mechanisms that produce the vulgar belief make it a nonstarter, as far as justification goes. These objections seem to find support from Hume's own disparaging remark that "trivial qualities of the fancy, conducted by . . . false suppositions" produce belief in body (T 1.4.2.56). My positive case for the defeasible justification of the vulgar belief in body is simple, though it builds on my account of philosophical justification presented in the last chapter. The vulgar belief in body is defeasibly justified because it is produced by permanent, irresistible, and universal propensities. I offer four lines of textual evidence in support of this claim.

First, Hume insists that even those philosophers who have reasoned themselves out of the vulgar version of belief in body still retain a permanent and irresistible propensity toward the belief in body (T 1.4.2.46–55). In fact, the

2. Thomas Reid is an early representative of this tradition. According to Reid, Hume demands arguments in support of all justified beliefs in contingent matters of fact beyond our own consciousness. Since there are no such arguments for belief in bodies, such belief is not justified. "Supposing certain impressions and ideas to exist in my mind, I cannot, from their existence, infer the existence of anything else: my impressions and ideas are the only existences of which I can have any knowledge" (Reid, *Inquiry*, 96). Later Reid comments: "I think it is evident, that we cannot, by reasoning from our sensations, collect the existence of bodies at all, far less any of their qualities. This hath been proved by unanswerable arguments by the Bishop of Cloyne, and by the author of the 'Treatise of Human Nature.'" Reid, *Inquiry*, 122. Reid chides Hume for inconsistency because Hume does not require arguments to justify introspective belief in impressions and ideas, as well as perceptual belief in bodies. Reid, *Inquiry*, 129–30. Contemporary epistemologists in the Reidian tradition continue to perpetuate this interpretation of Hume. The moral of Hume's external world skepticism, according to the Reidians, is that his justificatory standards are far too high; he demands arguments for beliefs that are properly basic. Plantinga, *Warrant*, 84–85; Alston, *The Reliability of Sense Perception* (Ithaca: Cornell University Press, 1993), 126–29; *Beyond "Justification,"* 215; Greco, "Skepticism About the External World," in *The Oxford Handbook of Skepticism*, ed. John Greco (New York: Oxford University Press, 2008), 111–13, 121–24; "Virtues," 305–8. Recent Hume scholars have argued that the vulgar belief is not justified because the propensities that produce the vulgar belief are not rational (Stroud, *Hume*, 109–110; Fogelin, *Hume's Skepticism*, 66, 78; Winkler, "Hume on Scepticism," 142–58), or not reliable (Schmitt, *Hume's Epistemology*, 259–63, 268–71, 274), or do not tend to produce stable beliefs (Loeb, *Stability*, 12–13, 139–62), or are contradictory (Annemarie Butler, "Vulgar Habits and Hume's Double Vision Argument," in *The Journal of Scottish Philosophy* 8.2 [2010], 170). Stefanie Rocknak could be added to this list. Rocknak follows Loeb in holding that only the products of custom-based causal reasoning are justified. Rocknak, *Imagined Causes: Hume's Conception of Objects* (New York: Springer, 2013), 231–32, 63–63; Loeb, *Stability*, 55–56. But the "'vulgar' belief is not a function of the relation of cause and effect, but rather of just the relation of resemblance." Rocknak, *Imagined*, 179.

original propensities which produce the vulgar belief are so strong that they prevent us from ever fully or permanently relinquishing it. To begin with, belief in the philosophical system of double existences (which supplants the vulgar belief in body) depends on the force of the natural propensities that produce the vulgar belief. After "a little reflection" destroys the vulgar belief "that our perceptions have a continu'd existence, by shewing that they have a dependent one, 'twou'd naturally be expected, that we must altogether reject the opinion, that there is such a thing in nature as a continu'd existence, which is preserv'd even when it no longer appears to the senses" (T 1.4.2.50). The rejection of all continued existences "is, in a manner, [the] necessary consequence" of rejecting the vulgar belief (T 1.4.2.50). But the psychological force of our natural, instinctive impulse

> can stop our progress, even in the midst of our most profound reflections, and keep us from running on with all the consequences of any philosophical opinion. Thus tho' we clearly perceive the dependence and interruption of our perceptions, we stop short in our carreer, and never upon that account reject the notion of an independent and continu'd existence. That opinion has taken such deep root in the imagination, that 'tis impossible ever to eradicate it, nor will any strain'd metaphysical conviction of the dependence of our perceptions be sufficient for that purpose. (T 1.4.2.51)

Furthermore, even philosophers who officially deny the vulgar belief quickly resume it again. "'Tis certain, that almost all mankind, and even philosophers themselves, for the greatest part of their lives" hold the vulgar belief (T 1.4.2.38). With the "least negligence or inattention" to reason, philosophers "can easily return to our vulgar and natural notions. Accordingly we find, that philosophers neglect not this advantage; but immediately upon leaving their closets, mingle with the rest of mankind in those exploded opinions, that our perceptions are our only objects, and continue identically and uninterruptedly the same in all their interrupted appearances" (T 1.4.2.53). In the final section of Book 1, Hume says that causal inferences and belief in "the continu'd existence of external objects, when absent from the senses" are "equally natural and necessary in the human mind" (T 1.4.7.4). The propensities to believe in body and to make causal inferences are equally permanent, irresistible, and universal, and therefore equally justifying.

Second, the vulgar belief's "primary recommendation . . . to . . . the imagination" gives it a justificatory advantage over the philosophical system of double existences (T 1.4.2.46). Hume criticizes the double existence theory not only because it "has no primary recommendation . . . to reason" but also because it has "no primary recommendation to the *imagination*" (T 1.4.2.46–48). This criticism presupposes that such a primary recommendation would go some way toward justifying the double existence theory. So far as justification goes, the vulgar belief is better off insofar as it has this primary recommendation to the imagination. Hume links the language of natural belief with that of "the primary recommendation" when he says "Tho' this opinion [the vulgar belief] be false, 'tis the most natural of any, and has alone any primary recommendation to the fancy" (T 1.4.2.48). My reading makes sense of this contrast between the vulgar belief and the double existence theory. The vulgar belief has a primary recommendation to the imagination because it is produced by permanent, irresistible, and universal propensities which are defeasibly justifying.[3] But the philosophical system of double existences derives defeasible justification neither from reason nor from the imagination.

Third, Hume indicates that the vulgar belief in body is produced by permanent, irresistible, and universal principles when he discusses it in Part 1 of Section 12 of the first *Enquiry*.[4]

> It seems evident, that men are carried, by a natural instinct or prepossession, to repose faith in their senses; and that, without any reasoning, or even almost before the use of reason, we always suppose an external universe, which depends not on our perception, but would exist, though we and every sensible creature were absent or annihilated. Even the animal creation are governed by a like opinion, and preserve this belief of external objects, in all their thoughts, designs and actions.
>
> It seems also evident, that, when men follow this blind and powerful instinct of nature, they always suppose the very images, presented by the senses, to be the external objects, and never entertain any suspicion, that the one are nothing but representations of the other. (EHU 12.7–8)

The first paragraph (12.7) points out that belief in an external universe is permanent, irresistible, and universal. This entails that it is produced by a permanent, irresistible, and universal propensity. The second paragraph (12.8) specifies that the PIU principle that produces the PIU belief in the external world, specifically produces the *vulgar* version of belief in the external world. That is, the PIU principle produces the belief that the very perceptions present to our senses continue to exist when they are no longer present to our minds.

---

3. An anonymous referee suggests an alternative reading of T 1.4.2.46–48. Perhaps Hume is arguing that the philosophical system is even worse off than the vulgar belief (although neither is defeasibly justified). On this reading, a primary recommendation to the imagination is better than nothing, but not sufficient for defeasible justification. While this reading of T 1.4.2.46–48 is possible, I think my reading is more probable in view of its coherence with the rest of Hume's theory and language of justification.

4. Hume has no general discussion of PIU principles in the *Enquiry* like he does in *Treatise* 1.4.4.1. I do not need to claim here that PIU principles have the same epistemic importance in the later work than in the *Treatise*. All I mean to point out here is that in the *Enquiry* too, Hume describes the vulgar belief as in fact produced by PIU principles.

Despite the important differences between the *Treatise* and the *Enquiry*, the consistency of Hume's view of the vulgar belief across both works on this point buttresses my reading. The first *Enquiry* does not explicitly say that philosophy endorses PIU principles; it has no passage parallel to *Treatise* 1.4.4.1–2. Moreover, the first *Enquiry* omits the *Treatise*'s detailed description of the mechanisms that produce the vulgar belief in body. But the *Enquiry* does agree with my reading of the *Treatise* in its general description of the vulgar belief as produced by PIU principles.

Fourth, in a 1751 letter to Gilbert Elliot of Minto, Hume makes the propensity to believe in body paradigmatic of a narrowly natural, defeasibly justified belief. Hume had sent Elliot an early draft of the *Dialogues Concerning Natural Religion*. In his letter, he identifies a crucial issue for one of Cleanthes' theistic arguments. In Part 3 of the *Dialogues*, Cleanthes claims that even if the inference to a designer is, strictly speaking, improbable, it is natural, universal, and irresistible for anyone who reflects on the causes of the world's order. Therefore, even by Philo's own skeptical lights, the belief in a designer is defeasibly justified. Hume marks the strength of this argument with an interjection by the narrator: "Here I could observe, *Hermippus*, that *Philo* was a little embarrassed and confounded" and "hesitated in delivering an answer" (DNR 3.10). Philo (Hume's spokesman) takes Cleanthes' argument seriously just because he agrees that a narrowly natural belief is at least defeasibly justified.[5] In the letter to Elliot, Hume writes of this exchange that

> I cou'd wish that Cleanthes' Argument coud be so analys'd, as to be render'd quite formal and regular. The Propensity of the Mind towards it, unless that Propensity were as strong and universal as that to believe in our Senses and Experience, will still, I am afraid, be esteem'd a suspicious Foundation. Tis here I wish for your Assistance. We must endeavour to prove that this Propensity is somewhat different from our Inclination to find our own Figures in the Clouds, our Face in the Moon, our Passions and Sentiments even in inanimate Matter. Such an Inclination may, and ought to be controul'd, and can never be a legitimate Ground of Assent. (HL i. 155)

The propensity to believe in continued and distinct existences—that is, "to believe in our Senses and Experience"—is paradigmatic of those propensities which are "strong and universal" enough to justify their deliverances defeasibly.

One might object that Hume must not regard the vulgar belief as defeasibly justified because throughout his explanation of its formation he describes it in terms of "fictions" and mistakes. The vulgar make three mistakes, in fact.

---

5. I discuss Hume's arguments against this kind of justification for theism in greater detail in "Hume and the Implanted Knowledge of God," *Journal of Scottish Philosophy*, 13(1), 2015: 17–35.

First, their ascription of duration to any unchanging object is "only by a fiction of the imagination" (T 1.4.2.29; cf. 1.2.3.11, 1.2.5.29). Second, their identification of exactly resembling but interrupted perceptions is a mistake. We are subject to "the error and deception with regard to identity, when we attribute it to our resembling perceptions, notwithstanding their interruption" (T 1.4.2.32). Third, their supposition that exactly resembling but interrupted perceptions are the intermittent appearances of a single, continued existence is also a fiction. We have "a propension to unite these broken appearances by the fiction of a continu'd existence" (T 1.4.2.36).

The first mistake—the fictitious ascription of duration to an unchanging object—deserves its own treatment. In order to explain why the vulgar attribute identity to their interrupted perceptions, Hume first needs to explain how the idea of identity is acquired (T 1.4.2.26–30). Two conditions are essential to a self-identical item: it must be unchanging and it must have duration (T 1.1.5.4). However, by Hume's lights, duration proper only applies to successions of items, not to single, unchanging items. We apply the idea of duration to unchanging things only by way of a fiction (T 1.2.3.11; 1.2.5.29; 1.4.2.29). It follows that the idea of a self-identical thing is a contradictory fiction. On the plausible assumption that no plain contradiction is defeasibly justified (its falsehood is intuitively certain), the vulgar belief in the identity of interrupted perceptions is not defeasibly justified.

This argument proves too much because identity ascriptions are not unique to the vulgar belief in body. If Hume's argument against identity disproved the defeasible justification of *all* identity claims, then it would be the most powerful skeptical argument in his corpus by far. He would surely draw attention to so radical a claim—if he made it. But Hume makes no fuss about such a claim. Even commentators who take Hume to have a skeptical argument against identity point this out.[6] Furthermore, an argument against the intelligibility of identity would not just underwrite a radically skeptical position, it would also overturn Hume's previous account of identity.[7] But he gives no indication of recanting his first account of so fundamental a concept. To be sure, his position on duration and identity is a riddle that needs further attention.[8] But it is not plausible that he rejects the vulgar belief in body on the grounds that all identity claims are unjustified nonsense.

The second and third mistakes—attributing identity and continued existence to exactly resembling, intermittently appearing perceptions—can be treated together. The important point is that there is nothing contradictory or impossible about believing in the identity and continued existence of these

6. Fogelin, *Hume's Skepticism*, 73; Noonan, *Hume on Knowledge*, 179. It is worth noting that Hume never mentions identity in discussing the vulgar belief in the first *Enquiry*.

7. Kemp Smith, *Philosophy*, 475n1.

8. Cf. Baxter, *Hume's Difficulty*, and Rocknak, *Imagined Causes*.

perceptions, even though the beliefs happen to be false.[9] Hume thinks it is perfectly conceivable that perceptions enter, leave, and return to our minds, without change or interruption in their existence (T 1.4.2.38–40). "The supposition of the continu'd existence of sensible objects or perceptions involves no contradiction" (T 1.4.2.40). If the vulgar belief is not contradictory, then why does Hume call it a fiction? The absence of a supporting argument is no objection, since the belief is produced by justifying natural propensities. The only explanation for the language of "fiction" and "mistake" is that Hume is anticipating the results of the defeater argument that he proposes later (T 1.4.2.44–45).[10] This language foreshadows his conclusion that the vulgar belief is not justified overall. But it does not indicate that the vulgar belief lacks defeasible justification.

The Defeat of the Vulgar Belief in Body

Although the vulgar belief is defeasibly justified, it nonetheless faces a rebutting defeater that Hume explains in *Treatise* 1.4.2.45. Most commentators deny the defeasible justification of the vulgar belief. Consequently, they often either fail to recognize a distinct skeptical argument in *Treatise* 1.4.2.45 or else they treat it in a cursory fashion—as if it were simply the coup de grace that finishes a mortally wounded theory.[11] It is therefore worth quoting the passage in full:

> 'TWILL first be proper to observe a few of those experiments, which convince us, that our perceptions are not possest of any independent existence. When we press one eye with a finger, we immediately perceive all the objects to become double, and one half of them to be remov'd from their common and natural

9. Cf. Annemarie Butler, "Hume on Believing the Vulgar Fiction of Continued Existence," *History of Philosophy Quarterly* 27.3, 242–45.

10. As I noted in chapter 2, Hume never uses the language of "defeaters." But the concept of defeat is present: evidence which renders an otherwise epistemically blameless belief, unacceptable.

11. Mounce and Fogelin do not even note the defeater argument. Mounce, *Hume's Naturalism*, 53–58; Fogelin, *Hume's Skepticism*, 64–92. Garrett briefly notes the argument, but does not regard it as skeptical in the epistemological sense. Garrett, *Cognition*, 211. Several commentators take very brief note of it after extended discussions of what they regard as more serious defects in the vulgar belief. Stroud, *Hume*, 110–11; Loeb, *Stability*, 152; Noonan, *Hume on Knowledge*, 182; Winkler, "Hume on Scepticism," 154–55. Fogelin does note the significance of *Treatise* 1.4.2.45 in his later work. Fogelin, "Hume's Skepticism," 227; *Hume's Skeptical Crisis*, 58–59, 79. Broughton seems to recognize that T 1.4.2.45 contains Hume's crucial skeptical argument against belief in body. Broughton, "Hume's Naturalism," 430. Schmitt recognizes that Hume gives no basis for denying the vulgar belief other than the argument of T 1.4.2.45. Schmitt, *Hume's Epistemology*, 266. Schmitt goes on to argue that the falsity of the vulgar belief leads Hume to reject the reliability of the propensity that produces this belief, which in turn leads him to conclude that the belief is not even defeasibly justified. Ibid., 268–69. Schmitt's reliabilist interpretation leads him to assess the epistemic significance of the argument differently than I do.

> position. But as we do not attribute a continu'd existence to both these perceptions, and as they are both of the same nature, we clearly perceive, that all our perceptions are dependent on our organs, and the disposition of our nerves and animal spirits. This opinion is confirm'd by the seeming encrease and diminution of objects, according to their distance; by the apparent alterations in their figure; by the changes in their colour and other qualities from our sickness and distempers; and by an infinite number of other experiments of the same kind; from all which we learn, that our sensible perceptions are not possest of any distinct or independent existence. (T 1.4.2.45)

Hume's argument runs like this (filling in some implicit premises):

1. If our perceptions are causally independent of our sensing them, then they continue to exist when not present to our senses (T 1.4.2.2).[12]
2. But some perceptions do not continue to exist when they are not present to our senses ("we do not attribute a continu'd existence to both these perceptions," T 1.4.2.45).
3. Therefore, some perceptions causally depend on our sensing them. (From 1 and 2.)
4. But all of our sensory perceptions are alike ("they are both of the same nature," T 1.4.2.45).
5. Like effects have like causes (fourth rule of causal reasoning, T 1.3.15.6).
6. Therefore, all of our sense perceptions have the same causes. (From 4 and 5.)
7. Therefore, all sense perceptions depend for their existence on our sensing them. (From 3 and 6.)

Whether this argument is sound is not important for present purposes.[13] The relevant point is that Hume thinks it is sound: "The natural consequence of this reasoning shou'd be, that our perceptions have no more a continu'd than an independent existence" (T 1.4.2.46). He draws the conclusion that none of our perceptions endure when they vanish from our senses. The problem with the vulgar belief is not the lack of supporting evidence but the presence of falsifying evidence (in a causal or inductive argument).

---

12. Hume asserts a biconditional relationship between continued existence and causal independence: "For if the objects of our senses continue to exist, even when they are not perceiv'd, their existence is of course independent of and distinct from the perception; and *vice versa*, if their existence be independent of the perception and distinct from it, they must continue to exist, even tho' they be not perceiv'd" (T 1.4.2.2).

13. For critiques, see Stroud, *Hume*, 111, and Schmitt, *Hume's Epistemology*, 272–73. Fogelin simply says "this is little more than a gesture in the direction of an argument." Fogelin, *Hume's Skeptical Crisis*, 59.

THE SYSTEM OF DOUBLE EXISTENCES

Reflection upon the causal argument against the continued, distinct existence of perceptions naturally leads "philosophers" to posit a system of double existences (T 1.4.2.46). The philosophers "distinguish betwixt perceptions and objects, of which the former are suppos'd to be interrupted, and perishing, and different at every different return; the latter to be uninterrupted, and to preserve a continu'd existence and identity" (T 1.4.2.46).[14] Natural principles also produce two other subsidiary features of the philosophical system. First, the philosophers assume that objects resemble mere perceptions (T 1.4.2.54). (By "mere perceptions" I mean those perceptions which are present to our senses, in contrast to that supposed class of perceptions which are bodies). Second, the philosophers assume that each perception is caused by an object that resembles it (T 1.4.2.55).

Hume clearly indicates the negative epistemic status of the philosophical system of double existences. He emphasizes that it "*has no primary recommendation either to reason or the imagination*" (T 1.4.2.46). On the contrary, it "*acquires all its influence on the imagination*" from the vulgar system (T 1.4.2.46). The philosophical system is "liable to the same difficulties" as the vulgar system, "and is over-and-above loaded with this absurdity, that it at once denies and establishes the vulgar supposition" (T 1.4.2.56). The philosophers, however, "arbitrarily invent a new set of perceptions," never present to any mind, which they construe as objects (T 1.4.2.56). The arbitrariness of positing of a new set of perceptions to serve as objects (that is, as continued and distinct existences) further vitiates the credibility of the philosophical system of double existences.

At the close of *Treatise* 1.4.2, Hume concludes that the vulgar and philosophical beliefs in body comprise a "confusion of groundless and extraordinary opinions" (T 1.4.2.56). Hume intends *Treatise* 1.4.2 (concerning the senses) to have the same negative epistemic result as *Treatise* 1.4.1 (concerning the understanding): "'Tis impossible upon any system to defend

14. Loeb mistakenly asserts that the philosophical system of double existences posits objects which are "specifically different" from perceptions, so that the philosophical system is unintelligible. Loeb, *Stability*, 163–64. Hume clearly asserts just the opposite. Already Hume has affirmed that "'tis impossible for us so much as to conceive or form an idea of any thing specifically different from ideas and impressions" (T 1.2.6.8). So Hume says that the adherents of the system of double existences "arbitrarily invent a new set of perceptions, to which they attribute these qualities. I say, a new set of perceptions: For we may well suppose in general, but 'tis impossible for us distinctly to conceive, objects to be in their nature any thing but exactly the same with perceptions" (T 1.4.2.56). Hume does not say that the system of double existences involves an inconceivable supposition, but an arbitrary and unsupportable supposition. He insists that we construe the system of double existences as an intelligible position, and the only way to do this is to take its "objects" as a postulated set of perceptions, rather than as items "specifically different" from perceptions. Winkler proves this point decisively. Winkler, "Hume on Scepticism and the Senses," 136–39.

either our understanding or senses; and we but expose them farther when we endeavour to justify them in that manner" (T 1.4.2.57). Although epistemologically defeated, these beliefs are nonetheless psychologically incorrigible:

> Carelessness and in-attention alone can afford us any remedy. For this reason I rely entirely upon them; and take it for granted, whatever may be the reader's opinion at this present moment, that an hour hence he will be persuaded there is both an external and internal world. (T 1.4.2.57)

But Hume is not even done criticizing the system of double existences.

In *Treatise* 1.4.4, Hume presents a rebutting defeater for the double existence theory. A good causal inference leads from the system of double existences to the belief that continued and distinct existences have only primary qualities, not the secondary qualities that appear to our senses (T 1.4.4.3–5). This refutes one of the claims of the system of double existences, the claim that our mere perceptions are caused by resembling objects, objects with the same qualities as our mere perceptions. The defeater argument forces the philosopher to retrench from the double existence theory to the modern philosophy.

The argument from the system of double existences to the modern philosophy is "that deriv'd from the variations of those impressions, even while the external object, to all appearance, continues the same" (T 1.4.4.3). We might formalize the argument of *Treatise* 1.4.4.4 as follows:

1. "The same object cannot, at the same time, be endow'd with different qualities of the same sense." (For example, a cloud cannot simultaneously be white and orange in the same place).
2. "[T]he same quality cannot resemble impressions entirely different." (For example, no quality can resemble both a white and an orange color-impression).
3. Successive impressions from the same sense modality, which we suppose are caused by one unchanging object, are entirely different.[15] (For example, in the course of a sunset, we successively experience white and orange color-impressions which we presume are caused by an unchanging cloud. The white and orange color-impressions are entirely different from each other).

15. Hume gives many examples to justify premise 3. We have incompatible impressions supposedly caused by a single unchanging object under the following circumstances: "Upon the different situations of our health: A man in a malady feels a disagreeable taste in meats, which before pleas'd him the most. Upon the different complexions and constitutions of men: That seems bitter to one, which is sweet to another. Upon the difference of their external situation and position: Colours reflected from the clouds change according to the distance of the clouds, and according to the angle they make with the eye and luminous body. Fire also communicates the sensation of pleasure at one distance, and that of pain at another. Instances of this kind are very numerous and frequent" (T 1.4.4.3).

4. The qualities of one unchanging object cannot resemble all of the entirely different impressions which it supposedly causes. (From 2 and 3). (For example, the cloud cannot possess a quality which resembles both the white and orange color-impressions which we presume it causes).
5. Either objects change their qualities whenever our sense perceptions of them change, or else they cause sense perceptions which they do not resemble.
6. But objects do not change whenever our sense perceptions of them change.[16]
7. So objects cause sense perceptions which they do not resemble. (From 5 and 6).
8. "Now from like effects we presume like causes."
9. The impressions which are not caused by a resembling object "are in appearance nothing different from the other impressions of colour, sound, &c."
10. "We conclude, therefore, that they are, all of them, deriv'd from a like origin"—that is, we conclude that all (secondary) impressions have non-resembling causes. (From 8 and 9).

Hume goes on to explain how the claims of the "modern philosophy" then follow: "For upon the removal of sounds, colours, heat, cold, and other sensible qualities, from the rank of continu'd independent existences, we are reduc'd merely to what are call'd primary qualities, as the only *real* ones, of which we have any adequate notion" (T 1.4.4.5).

For my purposes, the actual soundness of this argument is irrelevant. Hume treats it as valid. On my reading Hume himself accepts all of the premises except premise 6, which assumes the existence of enduring mind-independent objects. The adherent of the double existence theory, however, accepts premise 6, and presumably would accept the other premises as well. Hume regards this as a successful *reductio* skeptical argument against anyone who starts from the assumption that the double existence theory is true.

THE MODERN PHILOSOPHY

The modern philosophy lacks even defeasible justification, and it also faces a crushing defeater. The modern philosophy arises only as a refinement of the double existence theory. But the double existence theory itself lacks even defeasible justification. So the modern philosophy arises only as a refinement

16. The assumption here seems to be that if objects change with our sense perceptions of them, then no enduring mind-independent objects exist.

on a theory that was already dead on arrival—a refinement that does nothing to fix the epistemic flaws of its progenitor. Thus the modern theory has no primary recommendation either to reason or to the imagination, and is not defeasibly justified.

The modern philosophy faces a further death blow. Hume argues (à la Berkeley) that primary qualities without secondary qualities are inconceivable. Therefore, belief in bodies with only primary qualities is necessarily false (T 1.4.4.6–15; cf. EHU 12.15).

> I assert, that instead of explaining the operations of external objects by its means, we utterly annihilate all these objects, and reduce ourselves to the opinions of the most extravagant scepticism concerning them. If colours, sounds, tastes, and smells be merely perceptions, nothing we can conceive of is possest of a real, continu'd, and independent existence; not even motion, extension, and solidity, which are the primary qualities chiefly insisted on. (T 1.4.4.6)

Hume's argument from the primary/secondary quality distinction to the unintelligibility of matter runs as follows. He argues that apart from the secondary qualities, we have no concept of the primary qualities of motion, extension, and solidity, and without either the secondary or primary qualities, we have no concept of objects. Our idea of motion depends upon our idea of an object that moves; an object in turn must be defined exclusively in terms of extension and solidity (T 1.4.4.7). But our concept of extension depends upon our concepts of color or solidity (T 1.4.4.8). So the idea of motion, which depends upon the idea of objects, must derive from the concept of solidity. But the idea of solidity itself also depends upon the idea of objects, so it cannot provide materials for defining objects (T 1.4.4.9). Without relying upon the secondary qualities, or the primary qualities of motion, extension, or solidity, we have no more materials with which to define objects. "Our modern philosophy, therefore, leaves us no just nor satisfactory idea of solidity; nor consequently of matter" (T 1.4.4.9).

THE ANTIENT PHILOSOPHY

One version of belief in body, "the antient philosophy," stands somewhat alone (T 1.4.3). The other three versions of belief in body naturally succeed one another. The vulgar belief is universally instinctive; its defeat gives rise the double existence theory; the defeat of the double existence theory yields the modern philosophy, which is the end of the line. But the "antient philosophy" does not fit into this genetic series of beliefs. It arises rather from conceptual confusion. The "Peripatetic" philosophers account for the changing qualities of allegedly identical objects by positing an unintelligible substratrum in which those qualities inhere. Hume bluntly asserts that what we think of as bodies are bundles of sensible qualities (T 1.4.3.2). The claims that these bundles are simple (that the qualities are identical with one another) at a time, and that changing bundles are identical over time, are evident contradictions. "The antient philosophy" is not even defeasibly justified. Examination of the basic ideas involved shows the intuitive falsehood of the substance metaphysic, so it lacks even defeasible justification.

THREE ADVANTAGES OF THIS READING

My claim is that the vulgar belief in body is defeasibly justified, but that (like all forms of belief in body) it suffers rational defeat and is philosophically unjustified—all things considered. My reading sheds light on the role that skepticism about body plays in Hume's broader skeptical crisis. It has at least three advantages over alternative interpretations.

First, on my reading, Hume's skepticism about the senses mirrors his skepticism about causal reasoning in the *Treatise*.[17] When he points out that causal inferences are founded on natural instinct rather than on any good argument, he draws no skeptical conclusion about causal reasoning (T 1.3.6). Thus far, on the contrary, philosophy sanctions good causal reasoning as defeasibly justifying (T 1.3.11–13, 1.3.15, 1.4.4.1). The skeptical problem arises because iterative higher-order reasoning itself generates undermining defeaters for any rationally inferred conclusion (T 1.4.1). The vulgar belief in body is also founded on natural instinct rather than any possible argument. But since philosophy sanctions narrowly natural instinct (T 1.4.4.1–2), the vulgar belief is defeasibly justified so far. The skeptical problem arises only because the vulgar belief faces a defeater from reason (T 1.4.2.45; cf. EHU 12.9). This symmetry suggests that we are tracking Hume's intentions correctly. The titles alone of *Treatise* 1.4.1 ("Of scepticism with regard to reason") and of 1.4.2 ("Of scepticism with regard to the senses") lead us to expect the symmetry.

Second, if it is true, as many interpreters hold, that the Title Principle plays a central role in resolving Hume's skeptical problems, then my reading explains how it can. The deliverances of reason and belief in body both face defeat from reason (T 1.4.1–2). But the Title Principle says

> Where reason is lively, and mixes itself with some propensity, it ought to be assented to. Where it does not, it never can have any title to operate upon us. (T 1.4.7.11)

17. Cf. Ainslie, *Hume's True Scepticism*, 108

The Title Principle permits us to ignore reason, if it is not lively or accompanied by a propensity. We have no propensity to assent to the rational defeaters to belief in reason or to belief in bodies. So the Title Principle deals with these defeaters in one stroke.

On the other hand, unless belief in body is defeasibly justified at the outset, then the Title Principle does not permit us to hold this belief even while Hume's skeptical crisis is easing up. The Title Principle says only that we may hold beliefs produced by reason, plus a certain propensity. But reason does not produce the vulgar or any other form of belief in body. So the Title Principle gives us no way to evaluate this belief. If we construe the Title Principle as giving necessary conditions for permissible belief (we may hold a belief a belief *only* if it is produced by reason and we have the right propensity to assent), then belief in body is impermissible even after Hume regains his "serious good-humour'd disposition" (T 1.4.7.11). So the Title Principle can restore belief in body to us in the end only if that belief is defeasibly justified in the beginning.

Third, the essential points of the *Treatise's* arguments about belief in body are reiterated in Part 1 of Section 12 of the first *Enquiry*. When Hume repeats the argument against the vulgar belief in body in the first *Enquiry*, he entirely omits any account of the propensities that produce it. The point emphasized is that the vulgar belief is produced by a powerful instinct of nature. However, Hume produces a rebutting defeater argument from reason against this natural belief. If the problem with the vulgar belief in the *Treatise* lies in the specific propensities that produce it, this is not the position of the *Enquiry*. There Hume relies entirely on the defeater argument to support his skepticism about body. Maybe Hume changed his mind, or maybe he omitted the analysis and critique of the source propensities just to shorten the chapter. In any case, my reading ascribes to Hume a unified view consistent in both works. In both he regards the source-propensities of the vulgar belief as narrowly natural and defeasibly justifying; his only objection to the vulgar belief is that it faces rational defeat. I return to this topic in chapter 7.

Reidian misreadings of Hume's external world skepticism are mistaken. Mere insistence on a broader foundationalism or nature's prerogatives does not overcome Hume's consequent skepticism about the external world. Hume is no rabid rationalist, demanding arguments where only natural instincts are necessary or sufficient. He happily concedes the justifying power of permanent, irresistible, and universal cognitive mechanisms. Like Reid, he grants that no belief is more natural than perceptual belief in bodies. But Hume holds that natural beliefs are defeasible even when justified. The vulgar belief in continued and distinct existences (as well as philosophical variants of this belief) face clear defeat from reason. This is Hume's real skeptical challenge.

Chapter Five

Hume's Purely Practical Response to Skepticism in the *Treatise*

Many non-skeptical interpreters of the *Treatise* identify the Title Principle as Hume's final epistemic norm. Don Garrett, the most prominent of these interpreters, gave the Title Principle its name.[1] In the midst of his complex response to the skeptical dilemma, Hume says that "Where reason is lively, and mixes itself with some propensity, it ought to be assented to. Where it does not, it never can have any title to operate upon us" (T 1.4.7.11). In Hume's main skeptical arguments, reason is not lively and does not mix itself with some propensity. So these skeptical arguments have no title to operate upon us. Some commentators, especially in the wake of Michael Ridge's "Epistemology Moralized" (2003), hold that the Title Principle is an epistemic norm with practical justification. This makes good sense out of the heavy emphasis on agreeability and utility in *Treatise* 1.4.7, but as I mentioned in the Introduction, it collapses Hume's distinction between practical and epistemic justification.[2] Garrett, on the other hand, argues that in *Treatise* 1.4.7, Hume's defense of the justification of beliefs formed in accordance with the Title Principle is purely epistemic.[3] It does not rest on agreeability and usefulness but on the alethic considerations of truth and probable truth; the practical benefits of following the Title Principle are real but epistemologically incidental. This sort of reading explains how Hume moves on with science and avoids conflating morality with epistemology. However, it does not give a satisfying explanation of why Hume lays so much stress on practical rather than alethic considerations in the latter half of *Treatise* 1.4.7.

In this chapter, I argue that Hume's response to his skeptical problem is not epistemic, but purely practical. First, I reexamine the crisis of *Treatise* 1.4.7,

1. Garrett, *Cognition*, 234–37.
2. Qu, "Hume's Practically Epistemic Conclusions?" 509–523.
3. Garrett, *Hume*, 227–37.

giving special attention to "philosophy," Hume's terminology for epistemic normativity. Hume faces a "life-or-philosophy" dilemma: due to his skeptical arguments, core beliefs (that is, practically indispensable beliefs of common life and science) are not philosophically acceptable. The Title Principle is not a philosophical norm but rather subordinates philosophical norms to practical interests.[4] Second, I explain Hume's practical justification for a moderate pursuit of philosophy. He has purely practical reasons for ignoring the skeptical demands of philosophy, and purely practical reasons for following philosophy in his constructive scientific research. Third, I argue that only the purely practical reading can explain the role of *reductio* skeptical arguments (at T 1.4.7.3 and T 1.4.7.5) that Hume's "philosophy" does not support and that the Title Principle does not answer.

This purely practical reading of Hume's response to skepticism has at least three advantages over its competitors. First, since it starts from Hume's own terminology of "philosophy" rather than the anachronistic terminology of "epistemology," it is more directly grounded in the text. Second, it makes better sense than the purely epistemic reading out of Hume's emphasis on agreeability and usefulness in *Treatise* 1.4.7 while maintaining his distinction between moral norms and norms that control scientific inquiry. Third, it gives a satisfying explanation of the *reductio* skeptical arguments: Hume wants to drive readers into skepticism by any means available, so that they will be motivated to accept his purely practical solution.[5]

THE SKEPTICAL CHALLENGE AND THE LIFE-OR-PHILOSOPHY DILEMMA

Treatise 1.4.7, "Conclusion of this Book," presents a final catalog of skeptical concerns (T 1.4.7.1–7) that provoke a crisis (T 1.4.7.8) followed by a resolution (T 1.4.7.9–15) that permits Hume to carry on with the science of man in Books 2 and 3. The most important skeptical arguments in the catalog are the arguments against the deliverances of the senses and reason that he has

4. At a general level, my reading agrees with Kevin Meeker, who construes the Title Principle as a pragmatic principle and not as an epistemic principle. Meeker, *Hume's Radical Scepticism and the Fate of Naturalized Epistemology* (New York: Palgrave Macmillan, 2013), 73–81. We arrive at this conclusion by different routes, however. Perhaps most importantly, I do not share Meeker's view that all actual beliefs are adjudged as equally justified under the Title Principle. Peter Millican also denies that Hume thinks the Title Principle provides an epistemic solution to his skeptical dilemma; that dilemma is insoluble in the *Treatise*. Millican, "Hume's Chief Argument," 96–98, 105n42.

5. As I explain in the Introduction, by *reductio* skeptical arguments I mean arguments intended to prove that on an opponent's assumptions, certain important beliefs are epistemologically unjustified. The aim of these arguments is not to make the opponent give up the premise assumptions, but to admit that the allegedly important belief is not epistemologically justified.

already made at length earlier in Part 4. Hume's skeptical arguments show that philosophy requires the suspension of reason-based beliefs (T 1.4.1, 1.4.7.6–7) and the suspension of belief in continued and distinct objects (T 1.4.2, 1.4.4, 1.4.7.4). The faculty of reason, recursively applied to its own deliverances, generates undermining defeaters for any reason-based belief. Reason also generates rebutting defeaters for the belief in body in both its vulgar form (T 1.4.2.45) and in its ostensibly philosophical form (T 1.4.4).

The rational defeat of Hume's core beliefs forces him, in the first half of "Conclusion of this book," to ask what maxim he should adopt to guide belief formation. Either he must refuse to assent to reason, or else he must give up core beliefs which are practically indispensable for common life (as well as for scientific research). He concludes "For my part, I know not what ought to be done in the present case" (T 1.4.7.7). Four paragraphs later, Hume adopts the Title Principle as the maxim to guide belief formation, a principle which indulges the occasional rejection of reason, particularly in the case of these skeptical challenges.

There are two possible ways to interpret the normative question Hume asks and then finally answers with the Title Principle. The first interpretation takes Hume as a "particularist" about philosophical justification. The skeptical arguments bring to light a contradiction between his pretheoretical beliefs (i) that philosophy demands that reason ought to be assented to, and (ii) that belief in body and positive reason-based beliefs are philosophically justified. Instead of giving up (ii), he searches for an adequate revision of (i), and at last happily lights upon the Title Principle. The Title Principle says that philosophy demands that we assent to reason if and only if it is lively and mixes with some propensity. We are therefore philosophically justified in dissenting from the skeptical arguments of *Treatise* 1.4.1–2 and retaining our core beliefs.[6]

---

6. In a very broad (perhaps unhelpfully broad) sense, all the interpreters who take the Title Principle to be Hume's final epistemic principle could be described as "particularist" interpreters, insofar as their Hume has (or finds) an epistemic principle that justifies his pretheoretical particular epistemic judgments. These interpreters include Kemp Smith, *Philosophy*, 131; Baier, *Progress*, 280; Garrett, *Cognition*, 234–35; Garrett, *Hume*, 227–37; Morris, "Hume's Conclusion," 109; Owen, *Hume's Reason*, 217; cf. 203n12; Ridge, "Epistemology Moralized," 189; Schmitt, *Hume's Epistemology*, 368–75. None of these interpreters represents Hume's line of thought so simplistically as I have characterized the particularist reading. They do not all regard Hume as pretheoretically committed to the epistemic principle that reason (taken in abstraction from the rest of human nature) ought always to be assented to, even if this is a principle against which he implicitly argues. Baier and Morris think that the skeptical defeat of our core beliefs is part of a reductio ad absurdum argument against others who might hold this kind of view. Garrett does describe the Title Principle as a reflective revision of his own default principle that "reason ought to be assented to." Garrett, *Hume*, 230.

In the rest of this section I defend the second interpretation, which takes Hume as a "methodist" about philosophical justification.[7] Faced with the contradiction between (i) and (ii), he gives up (ii) and retains (i). When Hume wonders "what ought to be done," he is not wondering what philosophy requires of him. The demands of philosophy are clear: assent to reason and give up your beliefs. He wonders rather what practical principle he should adopt toward philosophy, which will permit him to escape abject agnosticism without plunging him into utter credulity. On the methodist reading, the dangerous dilemma is a life-or-philosophy dilemma: a choice between, on one hand, retaining the core beliefs which make life and science practically possible, and on the other hand, adhering to the demands of philosophy and reason.

The life-or-philosophy dilemma emerges first in the form of a "manifest contradiction" at the end of Hume's discussion of belief in mind-independent objects (T 1.4.7.4). Natural and necessary though non-ratiocinative principles of the imagination lead us to believe in mind-independent objects. Equally natural and necessary principles of the imagination produce an inductive inference that contradicts the belief in mind-independent objects. The contradiction between reason and belief in body forces us to choose between the principles:

> How then shall we adjust those principles together? Which of them shall we prefer? Or in case we prefer neither of them, but successively assent to both, as is usual among philosophers, with what confidence can we afterwards usurp that glorious title, when we thus knowingly embrace a manifest contradiction? (T 1.4.7.4)

By preferring the non-ratiocinative principles which produce belief in external objects, we reject the deliverances of reason (in violation of the norms of philosophy). But by preferring reason we reject belief in external objects—which sabotages common life and scientific research. To vacillate, assenting to reason most of the time but to contrary principles when they produce belief in external objects, is unworthy of "philosophers." So either we must give up our belief in bodies, or violate the demands of philosophy.

The argument for "scepticism with regard to reason" (T 1.4.1) generates a "very dangerous dilemma" which again forces us to choose between adhering to reason and retaining our core beliefs (T 1.4.7.6–7). If we "reject all the

7. So far forth, I agree with Broughton that "Hume reaches his skeptical conclusion only because he cedes authority to several broad cognitive norms of clarity, coherence, and evidence. A different philosopher might have questioned the authority of these norms rather than accept such a negative outcome," but Hume "finds them to be in order as they stand, even though full reflection on the nature of the mind in light of these norms forces us to see that our most basic assumptions about the world do not deserve our assent." Broughton, "The Inquiry," 547.

trivial suggestions of the fancy and adhere to the understanding" (T 1.4.7.7), then we must suspend all the deliverances of demonstrative and probabilistic reason. Only a trivial propensity prevents us from rationally, recursively diminishing our confidence levels to nothing. But suspending all of the deliverances of reason is practically unlivable. On the other hand, we could depart from the demands of reason, in one of two different ways. We might "assent to every trivial suggestion of the fancy" and reject all of the deliverances of reason (T 1.4.7.6).[8] This option is intellectually outrageous: "if we assent to every trivial suggestion of the fancy; beside that these suggestions are often contrary to each other; they lead us into such errors, absurdities, and obscurities, that we must at last become asham'd of our credulity" (T 1.4.7.6). Alternatively, we might "establish it for a general maxim, that no refin'd or elaborate reasoning is ever to be receiv'd" (T 1.4.7.7). This option saves us from the self-subversion of reason without requiring us to accept every trivial suggestion of the fancy. But it still destroys all science and philosophy, which consist in refined reasoning. It lacks intellectual integrity on other grounds as well. The rejection of refined reasoning is justified by a chain of reasoning that is quite refined, and so the rejection is self-defeating. The maxim is also ad hoc: "You proceed upon one singular quality of the imagination [viz., the quality whereby we ignore refined reasoning], and by a parity of reason must embrace them all" (T 1.4.7.7). The unacceptability of the ban on refined reasoning means that we are again caught between the horns of "a false reason" (which bows to trivial suggestions of the fancy) "or none at all" (the suspension of the self-subverted deliverances of reason) (T 1.4.7.7). Hume concludes "For my part, I know not what ought to be done in the present case" (T 1.4.7.7).

Philosophy, as Hume construes it, unqualifiedly endorses assent to the deliverances of reason.[9] Hume describes good and bad inductive inferences specifically in terms of what the "philosophers" do and do not sanction (T 1.3.13.1). *Treatise* 1.4.7 bears this point out as well. To whatever extent we cut off "refin'd or elaborate reasoning," to that extent we "cut off entirely all science and philosophy" (T 1.4.7.7). The philosophers reject education and "trivial propensities of the imagination" like the propensity to project because they conflict with reason (T 1.3.9.19, 1.4.3.11–1.4.4.1). In short, philosophy

8. It seems that, for Hume, accepting all the trivial suggestions of the fancy implies rejecting all or virtually all of the deliverances of reason. This seems to be the implication of Hume's assertion that the rejection of the fancy entails adherence to the understanding. It also seems to be the implication of his previous assertion that the trivial propensities of the imagination are "opposite to the other principles of custom and reasoning" (T 1.4.4.1).

9. Garrett refers to the principle that "*reason ought to be assented to*" as a "default principle." Garrett, *Hume*, 230.

endorses the default principle that reason ought always to receive our assent. So the life-or-reason dilemma is also a life-or-philosophy dilemma.

The Title Principle, which permits us to ignore reason at points, expresses Hume's preference for "life" over "philosophy," when he is forced to choose. The Title Principle is not a philosophical norm; on the contrary, it violates the norms of philosophy. In the first place, the Title Principle recommends that we sometimes assent to reason, and sometimes to the conflicting principles. But to sometimes assent and sometimes dissent from reason is unworthy of the "glorious title" of philosophy (T 1.4.7.4). Furthermore, if "You proceed upon one singular quality of the imagination" then "by a parity of reason [you] must embrace all of them" (T 1.4.7.7). The Title Principle, by contrast, lets us reject reason in favor of trivial propensities in arbitrarily restricted circumstances.

Second, Hume consistently associates strict adherence to philosophy with total suspension of belief. If philosophy sanctioned the Title Principle, philosophy would not lead to suspension of our core beliefs, since the Title Principle permits us to ignore the key rational defeaters. Hume's first response to the "very refin'd and metaphysical" skeptical arguments he has been considering is to "reject all belief and reasoning" and "look upon no opinion even as more probable or likely than another" (T 1.4.7.8). He refers back to this moment as "*philosophical* melancholy and delirium"—a delirium characterized by unyielding commitment to the demands of philosophy and therefore of reason (T 1.4.7.9, emphasis mine). By contrast, it is when, in an anti-philosophical mood of spleen and indolence, he sloughs off these psychologically unsustainable demands that he regains his core beliefs (T 1.4.7.10). He resolves "never more to renounce the pleasures of life for the sake of reasoning and philosophy" (T 1.4.7.9). With "the returns of a serious good-humour'd disposition" Hume goes back to philosophy, though not to its skeptical demands (T 1.4.7.11). What changes with his mood is not his view of what philosophy requires (namely, suspension of core beliefs). What changes is the extent to which he is inclined to follow those requirements.

In later writings Hume stresses that philosophy is always on the side of the Pyrrhonian skeptic who prescribes the suspension of core beliefs. Summarizing the *Treatise*, he writes in the Abstract that "Philosophy would render us entirely *Pyrrhonian*, were not nature too strong for it" (AT 27). In the *Letter to a Gentleman* Hume identifies the Pyrrhonian moment in the *Treatise* with his "Philosophical Melancholy and Delusion," the moment of his closest adherence to philosophy's demands (L 21, 23). In the first *Enquiry* he insists that "Pyrrhonism, or the excessive principles of scepticism," cannot be refuted by philosophy but only by action (EHU 12.21). The excessive principles of skepticism "may flourish and triumph in the schools; where it is, indeed, difficult, if not impossible, to refute them" (EHU 12.21). The skeptic is in his "proper sphere" when he displays "those *philosophical* objections, which arise from more profound researches. Here he seems to have ample matter of triumph" (EHU 12.22). Skeptical objections show that mankind "are not able, by their most diligent enquiry, to satisfy themselves concerning the foundation of these operations, or to remove the objections, which may be raised against them" (EHU 12.23). Since philosophy leads to the very Pyrrhonism from which the Title Principle saves us, the Title Principle cannot itself belong to philosophy. Hume does not merely say that philosophy cannot refute Pyrrhonian skepticism, but that it positively supports Pyrrhonian skepticism by means of objections to our core beliefs. Furthermore, although the texts cited from the first *Enquiry* were published nine years after the *Treatise*, they are terminologically and materially continuous with the texts from the "Abstract" and the "Letter to a Gentleman." These earlier works in turn purport to give Hume's own explication of his position in the *Treatise*. It is therefore reasonable to look to Hume's later discussions of Pyrrhonism to illuminate the meaning of *Treatise* 1.4.7.

Third, the Title Principle cannot be a philosophical principle because of its relationship to the Inclination Principle. The Inclination Principle is what I call Hume's statement that "if we are philosophers, it ought only to be upon sceptical principles, and from an inclination, which we feel to the employing ourselves after that manner" (T 1.4.7.11). The statement immediately precedes the Title Principle. After cycling through melancholic and then splenetic attitudes toward philosophy, he at last comes to the following conclusion:

> Nay if we are philosophers, it ought only to be upon sceptical principles, and from an inclination, which we feel to the employing ourselves after that manner. Where reason is lively, and mixes itself with some propensity, it ought to be assented to. Where it does not, it never can have any title to operate upon us. (T 1.4.7.11)

The Inclination Principle says that we ought to pursue philosophy only from an inclination to do so. The Title Principle immediately follows the Inclination Principle, and clearly stands in an appositional relation to it. The text demands that we take the "ought" of the Inclination Principle and the "ought" of the Title Principle in the same sense; a shift in meanings would constitute a complete non sequitur. Either both principles, or neither of them, state norms internal to philosophy as a method of inquiry.

But the Inclination Principle cannot be a philosophical ought. It does not state a norm internal to philosophy, but rather a norm about when we should (and should not) follow philosophy itself. If the Inclination Principle were a philosophical norm, then philosophy would require us to sometimes disobey philosophical requirements—which is self-defeating.

Someone might object that the Inclination Principle, construed as a philosophical norm, is not necessarily self-defeating. We can see why not by way of an analogy. A utilitarian might hold that the maximization of happiness is the correct standard for moral action, but not always the correct motive for moral action. Perhaps we will maximize universal happiness most effectively if we simply try to love our families and be good friends, employees, and citizens. We can consistently hold both that (a) you should always *do* what maximizes universal happiness, and (b) you should not *try* to maximize universal happiness. Arguably, the Inclination Principle is a philosophical principle analogous to (b). The effect of not trying (or trying not) to follow philosophy when we feel disinclined to it may be that we will actually adhere to philosophy in spite of ourselves.

This analogy does not hold, however. The utilitarian position above crucially distinguishes between what we should do and what we should try to do. It does not say we should do and not do precisely the same thing. But the Inclination Principle is only about when we should and should not actually be philosophers, not about when we should and should not try to be philosophers. The same goes for the Title Principle. Philosophy says, "Reason ought always to be assented to," while the Title Principle says, "It is false that reason ought always to be assented to." To be consistent with the demands of philosophy, the Title Principle would have to say, "Where reason is not lively and does not mix with some propensity, it never can have any title to *motivate* us." Hume does not suggest that we will actually follow philosophy by trying to shirk it when disinclined, or that we will actually assent to reason by trying to ignore it sometimes.

Again, someone might suggest that the Title Principle is a philosophical principle because it is itself the product of philosophical reasoning. But not every product of the philosophical method of inquiry is ipso facto part of the method itself. The point at issue here is whether the Title Principle is one of the norms governing philosophy as a method of inquiry.

I conclude that the "ought" of neither the Inclination Principle nor of the Title Principle can be a philosophical "ought." Just as the Inclination Principle says, "Only follow philosophy when you feel like it," the Title Principle adds "And in particular, only follow reason, philosophy's chief authoritative faculty, when you feel like it." Both principles express Hume's permissive practical stance toward the unbending demands of philosophy as such.

THE PRACTICAL JUSTIFICATION FOR A MODERATE PURSUIT OF PHILOSOPHY

The Inclination Principle (and its corollary, the Title Principle) in fact express the main lesson of Hume's skeptical crisis. Since philosophy, consistently pursued, demands the suspension of our core beliefs, complete adherence to philosophy is no more practically justified than it is psychologically sustainable. However, the moderate pursuit of philosophy in the context of a mixed way of life affords pleasure and benefit to those who incline toward it. Philosophy provides a comparatively safer and more agreeable method of inquiry and belief formation than its main alternative, superstition. By Hume's lights, we morally approve of those characteristics which are agreeable or useful to their possessors or to others. The habit of moderately pursuing philosophy is therefore practically and morally justified.[10]

The Rejection of Total Adherence to Philosophy

Following philosophy to the point of suspending core beliefs is futile. Nature prevails over argument and we take up our defeated core beliefs again, whether we want to or not.

> HERE then I find myself absolutely and necessarily determin'd to live, and talk, and act like other people in the common affairs of life . . . my natural propensity, and the course of my animal spirits and passions reduce me to this indolent belief in the general maxims of the world. (T 1.4.7.10)

Many other texts support the claim that we are psychologically incapable of suspending our core beliefs for long. After giving the skeptical argument against reason, Hume emphasizes that "nature breaks the force of all sceptical arguments in time, and keeps them from having any considerable influence on the understanding" (T 1.4.1.12; cf. T 1.4.1.7). The same holds true with respect to skeptical arguments about the senses. The skeptic "must assent to the principle concerning the existence of body, tho' he cannot pretend by any arguments of philosophy to maintain its veracity" (T 1.4.2.1; cf. T 1.4.2.57). Likewise, the dangerous dilemma "is seldom or never thought of; and even where it has once been present to the mind, is quickly forgot, and leaves but a small impression behind it. Very refin'd reflections have little or no influ-

10. An anonymous reviewer objects that at this point in the *Treatise*, Hume has not yet staked out this moral theory, so he cannot be appealing to it here. But just because Hume's moral theory has not yet appeared in the *Treatise* does not necessarily mean that he does not hold it and even deploy it. At a minimum, a moderate pursuit of philosophy has those traits that Hume will later identify as those of which we morally approve.

ence upon us" (T 1.4.7.7). "Most fortunately it happens, that since reason is incapable of dispelling these clouds, nature herself suffices to that purpose, and cures me of this philosophical melancholy and delirium, either by relaxing this bent of mind, or by some avocation, and lively impression of my senses, which obliterate all these chimeras" (T 1.4.7.9). "I must yield to the current of nature, in submitting to my senses and understanding" (T 1.4.7.10). "Our author . . . upon the whole concludes, that we assent to our faculties, and employ our reason only because we cannot help it. Philosophy would render us entirely *Pyrrhonian*, were not nature too strong for it" (AT 27). "The great subverter of *Pyrrhonism* or the excessive principles of scepticism, is action, and employment, and the occupations of common life" (EHU 12.21). Pyrrhonism is not "durable" outside the philosopher's closet (EHU 12.23, cf. 12.24). In the *Letter to a Gentleman* Hume again notes that the Pyrrhonian doctrine is psychologically impossible to live by (L 19–20).

The mere fact that we have a natural and powerful psychological propensity to hold our core beliefs does not by itself practically justify our acceptance of them. Arguably, Hume makes this very point when he says "Very refin'd reflections have little or no influence upon us; and yet we do not, and cannot establish it for a rule, that they ought not to have any influence" (T 1.4.7.7). Consider, for example, someone who has a natural and powerful psychological propensity to believe that they can fly from the tops of tall buildings. This belief is ordinarily fatal to humans who act on it. It is practically rational to resist this belief for as long as possible, even if the belief is natural and ultimately irresistible. Hume explicitly says that he is not entirely opposed to resisting his natural inclinations. He simply demands that the benefit of resisting a natural belief outweigh the pain and trouble of the resistance: "Where I strive against my inclination, I shall have a good reason for my resistance" (T 1.4.7.10). By "good reason" Hume means a good practical reason, not a good theoretical reason. Theoretical reasons too obviously support the unnatural suspension of core beliefs. But the quoted sentence occurs as a response to preceding rhetorical questions about the practical value of torturing his brain "with subtilities and sophistries":

> Under what obligation do I lie of making such an abuse of time? And to what end can it serve either for the service of mankind, or for my own private interest? (T 1.4.7.10)

Faced with the life-or-philosophy dilemma, he resolves "never more to renounce the pleasures of life for the sake of reasoning and philosophy" (T 1.4.7.10).[11]

11. As I argue in chapter 7, Hume rejects Pyrrhonism on the same practical grounds in the first *Enquiry* (EHU 12.23). Pyrrhonism demands a psychologically impossible feat. Even if we managed to suspend our beliefs, only harm would result.

The Return to a Moderate Pursuit of Philosophy

When his "sentiments of spleen and indolence" pass and "a serious good-humour'd disposition" returns, Hume realizes that philosophy sometimes is pleasurable (T 1.4.7.11). In the right circumstances he feels "naturally *inclin'd*" to pursue philosophy (T 1.4.7.12). By resisting this inclination when it arises, "I *feel* I shou'd be a loser in point of pleasure; and this is the origin of my philosophy" (T 1.4.7.12).

Although Hume returns to the pursuit of philosophy, he does not return to the total suspension of his core beliefs which philosophy demands. He instead returns to a positive use of philosophy, deploying this normative method in belief formation and theory construction. "[I] am naturally *inclin'd*," he says, "to carry my view into all those subjects, about which I have met with so many disputes in the course of my reading and conversation" (T 1.4.7.12). He goes on to list all of the subjects about which he desires to form theories: moral good and evil, the foundations of government, the causes of the passions, aesthetics, truth and falsehood, reason and folly (T 1.4.7.12). One might object that Hume has no right to follow philosophy in his positive theory construction if he will not follow philosophy down the ineluctable path to Pyrrhonism. But Hume can reply that he is practically justified in making a positive use of philosophy, while ignoring its Pyrrhonian demands. He flouts philosophy's parity of reasoning requirement on practical grounds.

Hume flatly denies that all people should pursue philosophy. Not everyone is led, either by their strength or weakness of mind, to inquire beyond the bounds of common life (T 1.4.7.14). English gentlemen in particular frequently have this "earthy mixture." Hume sees no practical reason for "refining them into philosophers":

> Of such as these I pretend not to make philosophers, nor do I expect them either to be associates in these researches or auditors of these discoveries. They do well to keep themselves in their present situation. (T 1.4.7.14)

Philosophy does not hold intrinsic value for everyone. Some people, in some moods, enjoy pursuing philosophy to some extent. But those without the time or temperament for philosophy have no practical obligation to pursue it.

Hume concludes Book 1 by noting that just as he has a practical justification for holding epistemologically defeated beliefs, he also has a practical justification for holding them with certainty—at least from time to time (T 1.4.7.15). On reflection Hume holds all of his beliefs tentatively. But in the moment of belief-formation it is natural to form them with certainty, and it would cost too much psychological trouble to fight this natural tendency. Not only does Hume give practical justifications for his beliefs, but he also gives practical justifications for his confidence levels.

Against practical readings like mine, Donald Ainslie argues that if Hume is really an epistemic skeptic, he can have no practical reason to return to philosophy.[12] Ainslie asserts that "when Hume returns to philosophy, he continues to aim at truth."[13] If Hume accepts a negative epistemic verdict on his faculties, then he believes that it is impossible for him to attain knowledge of the truth by way of philosophy.[14] He therefore has no practical reason to continue to philosophize. Garrett makes a similar point when he says, "it is psychologically untenable, in [Hume's] psychology, to take pleasure in the satisfaction of either curiosity ('to know' foundations and principles) or ambition ('of contributing to the instruction of mankind, and of making a name by my inventions and discoveries') without taking one's own discoveries to be true or at least probably true."[15]

To begin with, Ainslie's objection does not adequately distinguish between Hume's beliefs at the time he decides to return to philosophy, and his beliefs as he practices philosophy. As Hume considers whether to return to philosophical enquiry, he believes that it can never produce philosophically justified beliefs. Ainslie is correct to point out the prima facie irrationality of a skeptic engaging in research. This, I take it, is precisely the force of the "notwithstanding" in Hume's statement that it is "proper we shou'd in general indulge our inclination in the most elaborate philosophical researches, notwithstanding our sceptical principles" (T 1.4.7.15). But Hume also believes that philosophical research will inevitably produce beliefs of which he will be, at least temporarily, subjectively certain. When we engage in enquiry, "assurance . . . always arises from an exact and full survey of an object" (T 1.4.7.15). Pursuing knowledge and seeming to attain it are agreeable and useful activities. When Hume decides to return to philosophizing, he foresees that he will not experience his labors as futile. On the contrary, he will seek and seem to attain truth, and find the whole process very agreeable. So he has good practical reason to return to philosophy, even if, at the time he makes this decision, he believes it will not actually produce justified beliefs.

A similar worry to Ainslie's is that my reading imputes to Hume two logically contradictory and psychologically incompatible beliefs.[16] On the back side of his skeptical crisis, Hume believes that the consistent pursuit of philosophy does not lead to true belief. He also believes that philosophical inquiry into particular questions will afford pleasure. But philosophical inquiry is likely to afford pleasure only if it is likely to lead to true belief. So Hume simultaneously believes that philosophy does and does not lead to true belief.

---

12. Ainslie, *Hume's True Scepticism*, 226–30.
13. Ibid., 228
14. Ibid., 230
15. Garrett, *Hume*, 232.
16. Thanks to an anonymous referee for raising this problem.

Holding these contradictory beliefs would make Hume a little bit crazy, if it is even psychologically possible to do so.

To answer this worry, I distinguish between occurrent beliefs and doxastic dispositions. My account does not necessarily saddle Hume with logically contradictory, simultaneously occurring beliefs. It does impute to him the simultaneous possession of contradictory doxastic dispositions. If and when Hume reflects on the outcome of consistent philosophical reasoning, he has an occurrent belief that (a) the consistent pursuit of philosophy does not lead to true belief. If and when Hume reflects on some particular question—for example, "What are the foundations of government?"—he does not have the occurrent belief (a). Instead, he has the more or less explicit belief that (b) philosophical inquiry into this question might very well lead to truth, thereby affording pleasure. Hume is disposed to assent to (a) when he ponders the second-order question about philosophy, and disposed to assent to (b) when he engages in first-order reflection about the world.

Is it psychologically possible to simultaneously possess these conflicting dispositions? I think it is. In fact, I think that philosophers often have similarly conflicting dispositions. Many of us actually do become (temporarily or permanently) convinced that some of our ordinary beliefs—say, in the reliability of one or all of our cognitive faculties, or in objective moral facts, or whatever—are probably false. When we are asked a second-order question about the reliability of these beliefs in the classroom or at a conference, we answer negatively. But as soon as we walk outside of our classrooms, we revert to making first-order judgments that presuppose the reliability of our faculties or the reality of objective morality. If we meet a student or colleague in the hall who reminds us of our official skepticism, we deny the first-order beliefs again, only to absent-mindedly resume them once more on the commute home. As Hume puts it, it is "usual among philosophers" to "successively assent" to conflicting doxastic principles, and "thus knowingly embrace a manifest contradiction" (T 1.4.7.4).

My account hinges on the fact that Hume's skepticism does not psychologically prevent him from holding beliefs when he returns to philosophical research.[17] He closes Book 1 of the *Treatise* by saying that despite his

17. The point is controversial. Garrett argues that for Hume, beliefs which we regard as epistemologically unjustified are not stable beliefs at all. Garrett, *Hume*, 232–33. The texts I cited earlier show that Hume does think that the skeptic will go on to hold stable beliefs which, at least on reflection, he admits are philosophically unjustified. Ainslie argues that Hume could pursue philosophy only insofar as he forgets his allegedly devastating skeptical arguments. I agree. But, Ainslie adds, throughout Books 2 and 3 of the *Treatise*, Hume recalls the very theories in Book 1 which give rise to his skeptical problems. He concludes that the skeptical arguments could not have been epistemically devastating after all. Ainslie, *Hume's True Scepticism*, 229–30. I do not think that the texts Ainslie cites shows that Hume simultaneously gives focused attention to his skeptical arguments and retains his confidence in his scientific beliefs.

skeptical principles he will inevitably experience and express certainty about the matters he investigates (T 1.4.7.15). When we engage in enquiry, "assurance . . . always arises from an exact and full survey of an object." When we yield to this propensity for assurance, we "are apt not only to forget our scepticism, but even our modesty too."[18]

The textual data already cited from *Treatise* 1.4.7 and the *Enquiry* makes it absolutely clear, I think, that Hume thinks that the epistemic skeptic has good practical reasons for returning to the core beliefs of common life and philosophy.[19] But *Dialogues* 1.9 must dispel any remaining doubt on this point. Philo (the skeptic) begins, "To whatever length any one may push his speculative principles of scepticism, he must act, I own, and live, and converse like other men; and for this conduct he is not obliged to give any other reason, than the absolute necessity he lies under of so doing" (DNR 1.9). The "absolute necessity" in view is both psychological and practical (cf. DNR 1.5–8). The skeptic is psychologically compelled to "act . . . and live, and converse like other men" (DNR 1.9). To act and converse like a non-skeptic is also practically necessary for life. As Cleanthes puts it, "If they [sceptics] be thoroughly in earnest, they will not long trouble the world with their doubts, cavils, and disputes" because they will fatally exit buildings by windows rather than doors (DNR 1.5). Even the most extreme Pyrrhonian skeptic who denies the epistemic justification of all beliefs whatsoever is practically justified in living and acting like other humans—that is, in living according to commonly received core beliefs of reason and the senses: "and for this conduct *he is not obliged to give any other reason*, than the absolute necessity he lies under of so doing" (emphasis mine). Psychological compulsion is a sufficient practical reason even for the extreme skeptic to return to common life.

Philo goes on to argue that if the skeptic is practically justified in returning to common life, then he is also practically justified in returning to philosophy:

> If he ever carries his speculations farther than this necessity constrains him, and philosophises, either on natural or moral subjects, he is allured by a certain pleasure and satisfaction, which he finds in employing himself after that manner. He considers besides, that every one, even in common life, is constrained to have

18. Schafer denies that Hume is a radical epistemic skeptic on the grounds that he appears to endorse various beliefs and principles on epistemic grounds in Book 1 of the *Treatise* and beyond. Schafer, "Curious Virtues," 8. But Hume explicitly warns that his use of terms of epistemic approbation such as "*'tis evident, 'tis certain, 'tis undeniable*" are "extorted from me by the present view of the object" and do not represent his considered judgement, and certainly not his ultimately skeptical principles (T 1.4.7.15). Hume only experiences any degree of confidence by turning his attention away from his skeptical arguments, not by neutralizing them.

19. In chapter 7 I argue that Hume also gives a purely practical response to skepticism in the *Enquiry*. However, at this stage, I am only claiming that in the *Enquiry*, the epistemic skeptic does have practical reasons for resuming core beliefs. That does not exclude the possibility that Hume might present an epistemic solution to skepticism in the *Enquiry* as well (although I think he does not).

> more or less of this philosophy; that from our earliest infancy we make continual advances in forming more general principles of conduct and reasoning; that the larger experience we acquire, and the stronger reason we are endued with, we always render our principles the more general and comprehensive; and that what we call *philosophy* is nothing but a more regular and methodical operation of the same kind. To philosophise on such subjects is nothing essentially different from reasoning on common life; and we may only expect greater stability, if not greater truth, from our philosophy, on account of its exacter and more scrupulous method of proceeding. (DNR 1.9)

Philo gives two practical justifications for the skeptic's return to philosophizing. The first is "a certain pleasure and satisfaction, which he finds in employing himself after that manner." The second ("He considers besides") is that the difference between reasoning in common life and in philosophy is only of one degree, not kind. If the skeptic is practically justified in returning to common life (and she is), then she is practically justified in philosophizing. I do not know how Hume could have given a more direct or explicit affirmative answer to the question, "Is the extreme epistemic skeptic practically justified in returning to philosophy?"

The Relative Practical Warrant of Philosophy and Superstition

Hume's subordination of philosophy to practical concerns could very well open the door to all sorts of intellectual bad behavior. A religionist or any other purveyor of absurdities could just as well say that their method of belief formation is as pleasant and useful as any other. In order to justify the pursuit of philosophy, Hume needs to show not only that philosophy has some practical benefit, but that it has more practical benefit than its rivals.[20]

He gives a practical criterion for the doxastic method we ought to prefer in our enquiries outside common life: "we ought only to deliberate concerning the choice of our guide, and ought to prefer that which is safest and most agreeable" (T 1.4.7.13). He divides the entire field of methodological alternatives for belief-formation outside the sphere of common life into two categories: philosophy and superstition. He argues that philosophy is the safest and most agreeable guide (T 1.4.7.13).

We may worry that Hume has no right to make claims about the practical consequences of suspending all belief, or of moderately pursuing philosophy, or of following superstition.[21] After all, these sorts of claims are based on

20. Ridge highlights this worry and provides his own answer. Ridge, "Epistemology Moralized," 167, 184–94. On Ridge's view, Hume himself finds reliance upon the understanding (rather than on superstition) immediately agreeable, and this gives him a practical justification for employing it. I am in basic sympathy with Ridge's line of argument.

21. Millican, "Hume's Chief Argument," 97–98.

inductive reasoning, which is philosophically unjustified. I think Hume would concede that these beliefs are indeed philosophically unjustified. But, as with other causal inferences, he holds them on the practical grounds that "it costs us too much pains to think otherwise" (T 1.4.7.11).[22] His beliefs about the practical consequences of belief-forming policies are themselves justified on practical grounds, and on practical grounds only.

THE ROLE OF *REDUCTIO* SKEPTICAL ARGUMENTS

The purely practical reading of Hume's response to skepticism can make excellent sense of the skeptical arguments in *Treatise* 1.4.7.3 and 1.4.7.5. These arguments are initially puzzling: measured by the standards of Hume's own philosophical norms and commitments, they completely fail. Furthermore, the Title Principle has no bearing on them at all. My explanation of them is that they are *reductio* arguments—that is, they are compelling skeptical problems for readers with different philosophical commitments than Hume. Once his readers feel trapped in a skeptical dilemma, he can offer them his purely practical solution: simply demur from the demands of philosophy on practical grounds when philosophy threatens core beliefs. The *reductio* arguments show such readers that they can adhere absolutely to their own philosophical standards only at the cost of giving up core beliefs which make life possible.

It is a unique advantage of the purely practical reading that it can explain these arguments. Some commentators simply neglect them.[23] Others deny that any of Hume's arguments successfully show that his core beliefs are epistemologically unjustified by his own lights.[24] This leveling strategy eliminates

---

22. An anonymous reviewer suggests that this response seems feeble. That may be so, but I think it is Hume's view. Furthermore, it is not clear to me why the response is feeble. The worry seems to be that Hume has no epistemic justification for some of his claims, though he ought to. But this worry is question-begging if, as I have argued, his entire position is that he is not obliged to have epistemic justification for all of his claims.

23. For example, Fogelin omits all discussion of *Treatise* 1.4.7.5. Fogelin, *Hume's Skepticism*. Owen focuses mainly on the argument in *Treatise* 1.4.1 and the dangerous dilemma to which it gives rise (T 1.4.7.7), but omits discussion of *Treatise* 1.4.7.3 or 1.4.7.5. Owen, *Hume's Reason*, especially chapter 9. Meeker focuses exclusively on *Treatise* 1.4.1 and its implications in *Hume's Radical Scepticism*, neglecting 1.4.7.3 and 1.4.7.5. Many more examples could be given.

24. For example, Garrett finds five skeptical (that is, doubt-inducing) considerations in *Treatise* 1.4.7 and treats them all as equally real contributors to Hume's skeptical crisis. Garrett, *Cognition*, 204–232; *Hume*, 218–27. Note however that for Garrett, Hume's skeptical crisis is not epistemic, but psychological. The skeptical considerations induce doubt, but Hume never asserts that, epistemologically, they ought to induce us to suspend our beliefs. Morris, like other *reductio* readers, distances Hume from all of the apparent skeptical threats in *Treatise* 1.4.7.3–7. Morris, "Hume's Conclusion," 89–110. These arise only for a "conscientious traditional metaphysician," not for Hume himself. Ibid 107. Morris also therefore gives all of Hume's skeptical considerations the same epistemic weight—nil. My reading actually resembles Morris and other *reductio* readings insofar as I think that some of Hume's skeptical threats only arise for projected interlocutors, not for himself. Against Morris

the need to give a special explanation for what I am calling *reductio* skeptical arguments. However, as I have argued above, the levelling strategy is unsustainable, since the arguments against reason and the senses do succeed, according to Hume.

Treatise 1.4.7.3

The first *reductio* arguments occur at the head of Hume's catalog of explanations for his frightened and confounded "forlorn solitude" and for his lack of confidence in further research (T 1.4.7.2–3).

> After the most accurate and exact of my reasonings, I can give no reason why I shou'd assent to it; and feel nothing but a *strong* propensity to consider objects *strongly* in that view, under which they appear to me. Experience is a principle, which instructs me in the several conjunctions of objects for the past. Habit is another principle, which determines me to expect the same for the future; and both of them conspiring to operate upon the imagination, make me form certain ideas in a more intense and lively manner, than others, which are not attended with the same advantages. Without this quality, by which the mind enlivens some ideas beyond others (which seemingly is so trivial, and so little founded on reason) we cou'd never assent to any argument, nor carry our view beyond those few objects, which are present to our senses. Nay, even to these objects we cou'd never attribute any existence, but what was dependent on the senses; and must comprehend them entirely in that succession of perceptions, which constitutes our self or person. Nay farther, even with relation to that succession, we cou'd only admit of those perceptions, which are immediately present to our consciousness, nor cou'd those lively images, with which the memory presents us, be ever receiv'd as true pictures of past perceptions. The memory, senses, and understanding are, therefore, all of them founded on the imagination, or the vivacity of ideas. (T 1.4.7.3)

In this passage, Hume points out that the vivifying propensity of the imagination which produces inferences is "trivial" and not "founded on reason." The passage seems to imply that in the absence of supporting argument, causal inferences lack epistemic justification.[25] Moreover, it implies that belief in continuing and distinct existences is unjustified since it is produced only

---

however I have argued that some of Hume's skeptical arguments against core beliefs do succeed by his own lights. Furthermore, unlike Morris and other *reductio* readings, I think that point of the *reductio* arguments is not to make the interlocutor give up on the premises that generate the argument, but to make the interlocutor give up on the claim that their core beliefs are philosophically justified.

25. Garrett has a different explanation of this conversational implicature. He says that Hume is simply expressing, in "strictly reportorial language," "a temporary feeling of diminished confidence in the three idea-enlivening mechanisms." Garrett, *Cognition*, 215. Note however that after the *Treatise*, Hume still manages to convey the suggestion of skepticism, even though he is not reporting his emotions in these texts.

by the imagination, not by argument. Even memory beliefs succumb to this criticism. Hume has previously made clear that beliefs based on causal inference and sensory beliefs are produced by PIU principles of the imagination; it follows that beliefs based on causal inference and sensory beliefs are defeasibly justified. He has everywhere taken the same for granted about memory, never hinting at any possible epistemic defect in it. The passage appears to represent an epistemic *volte-face*.[26]

Hume makes the same move in the "Abstract" of the *Treatise*.

> By all that has been said the reader will easily perceive, that the philosophy contained in this book is very sceptical, and tends to give us a notion of the imperfections and narrow limits of human understanding. Almost all reasoning is there reduced to experience; and the belief, which attends experience, is explained to be nothing but a peculiar sentiment, or lively conception produced by habit. Nor is this all. When we believe any thing of *external* existence, or suppose an object to exist a moment after it is no longer perceived, this belief is nothing but a sentiment of the same kind. Our author insists upon several other sceptical topics; and upon the whole concludes, that we assent to our faculties, and employ our reason only because we cannot help it. Philosophy would render us entirely *Pyrrhonian*, were not nature too strong for it. (AT 27)

Hume is plainly referring back to his analysis of causal inference as the example par excellence of the "very skeptical" nature of the philosophy of the *Treatise*. Beliefs produced by causal inference are simply feelings produced by habit, not reason. Belief in body is also a feeling not produced by reason or argument. The implication is that since reason does not produce these beliefs, they lack even defeasible epistemic justification, and that this is the source of Hume's skeptical predicament.[27]

But Hume cannot regard these skeptical threats as compelling by his own lights. He has already stated that philosophers approve of the vivifying principles of the imagination when they are permanent, universal, and irresistible (T 1.4.4.1). This statement comes after his full treatment of how these principles work. To suddenly reverse course and announce that causal inference and the senses do not defeasibly justify their products, just because they are caused by the unreasoning imagination, would be completely incoherent.

---

26. Singer notes how surprising it is that Hume should denigrate the epistemic status of the imagination, prior to citing any arguments that show its unreliability or weakness. Singer, "Hume's Extreme Skepticism," 599.

27. Schmitt takes Hume's argument in T 1.4.7.3 as genuinely skeptical, but construes the skeptical threat differently than I have. He takes T 1.4.7.3 (together with the first sentence of T 1.4.7.4) to indicate that "the Manifest Contradiction [in T 1.4.7.4] demonstrates the unreliability of imaginative operations and thereby raises a doubt about the reliability, hence about the defeasibly justifying power, of causal inference." Schmitt, *Hume's Epistemology*, 346. He says nothing in particular about the new worries about perceptual belief and memory.

Moreover, Hume nowhere else in his corpus calls into question the defeasible justification of memory beliefs. To do so would raise a skeptical challenge as profound as any other that he develops. If he seriously doubted the defeasible justification of memory beliefs we would have every reason to expect him to exploit this issue further and discuss it elsewhere. But he does not.

Finally, the Title Principle does not provide any obvious solution to the skeptical challenge against the senses and memory. The Title Principle endorses lively reason, but does not say anything explicitly about beliefs produced by the senses and memory.[28] So if the Title Principle is Hume's way of moving past the skeptical problems he takes seriously, then these skeptical problems are not among them.

A closer examination of *Treatise* 1.4.7.3 and Abstract 27 shows that Hume does not actually endorse the skeptical threats that he suggests. As Garrett points out about *Treatise* 1.4.7.3, Hume does not "argue or assert that (i) inductive conclusions, (ii) claims of continued and distinct existence, or (iii) memories are unworthy of belief."[29] He simply makes descriptive claims about our cognitive mechanisms. This description is prefaced by some forlorn questions:

> For with what confidence can I venture upon such bold enterprizes, when beside those numberless infirmities peculiar to myself, I find so many which are common to human nature? Can I be sure, that in leaving all establish'd opinions I am following truth; and by what criterion shall I distinguish her, even if fortune shou'd at last guide me on her foot-steps? (T 1.4.7.3)

These rhetorical questions invite the answers "With no confidence," and "No, I cannot be sure." They also invite the reader to take the rest of the paragraph as giving grounds for those negative verdicts. But Hume does not directly say that "Because the memory, senses, and understanding are founded on the vivacity of our ideas, therefore I have no grounds for trusting them."

Similarly, Abstract 27 does not explicitly reject causal inference and belief in body on the grounds that they are produced by the vivacity of ideas. Hume does give that impression, however, by sandwiching descriptive claims about these beliefs between statements of profound skepticism:

> By all that has been said [about causal inference] the reader will easily perceive, that the philosophy contained in this book is very sceptical, and tends to give us a notion of the imperfections and narrow limits of human understanding . . . Our author insists upon several other sceptical topics; and upon the whole concludes,

28. One possible response to this argument is that the Title Principle implicitly endorses whatever faculties contribute to lively reasoning. Since the senses and memory are necessary to lively reasoning, the Title Principle implicitly endorses them.

29. Garrett, *Cognition*, 214.

> that we assent to our faculties, and employ our reason only because we cannot help it. Philosophy would render us entirely *Pyrrhonian*, were not nature too strong for it. (AT 27)

However, this passage does not directly and explicitly identify the descriptive statements as the grounds for skepticism. Hume could—and he does—regard "other sceptical topics" as the basis for claiming that "Philosophy would render us entirely *Pyrrhonian*, were not nature too strong for it."

Treatise 1.4.7.5

In *Treatise* 1.4.7.5, Hume seems to suggest that his argument against the very idea of necessary connections between objects poses a skeptical challenge to a belief of some importance. The despairing tone of *Treatise* 1.4.7.5 is unmistakable. Hume's argument against objective necessary connections is presented as an example of the lack of "any degree of solidity and satisfaction in . . . our reasoning" (T 1.4.7.5). The argument seems "to turn into ridicule all our past pains and industry, and to discourage us from future enquiries" (ibid). It frustrates the most basic aims of science.

Hume certainly endorses the arguments against belief in objective necessary connections. In *Treatise* 1.3.14, he shows that we cannot even conceive of such a thing; objective necessary connections violate the Copy Principle. Furthermore, belief in objective necessary connections is produced by our propensity to project (T 1.3.14.25).[30] The philosophers disapprove of this propensity; its deliverances lack even defeasible justification (T 1.4.3.11–1.4.4.2). The propensity itself is easily suppressed, and careful reflection dislodges its deliverances from among our beliefs. Not even the Title Principle rescues this belief. The Title Principle endorses lively reason, but belief in objective necessary connections does not arise from reason at all. Hume alludes back to its dubious origins in *Treatise* 1.4.7.6 when he calls the belief an "illusion of the imagination" and a "trivial suggestion of the fancy."

Garrett construes *Treatise* 1.4.7.5 not so much as a direct epistemic skeptical threat to a core belief, but as a doubt-inducing consideration about reason in general: "This discovery presumably weighs against the probability that beliefs produced by reason are true by showing that the mind is subject to at least one pervasive illusion in the course of much or all of its most common

30. Hume signals that this propensity to project also leads us to think that sounds and smells have spatial location. In a footnote (T 1.3.14.25, footnote 32) he points forward to his discussion of this issue in T 1.4.5. But in T 1.4.5, Hume attributes the mistaken ascription of spatial location to what seems to be a distinct propensity—the propensity to add a relation (T 1.4.5.12). As Hume notes in T 1.4.5.12, the propensity to add a relation accounts for the representationalist belief that objects resemble the perceptions that they cause (cf. T 1.4.2.54–55).

and most important probable reasoning."[31] But it is odd that the falsehood of a belief *not* produced by reason should raise doubts about the reliability of beliefs that *are* produced by reason. It is certainly no surprise that beliefs never countenanced by philosophy anyway turn out to be fallacious, nor is it clear why that would raise a doubt about those beliefs that philosophy does countenance.

The reason why *Treatise* 1.4.7.5 cannot be taken seriously as a skeptical argument is that for Hume, the belief in objective necessary connections is completely dispensable; it is not a core belief.[32] It may be widely held, but it is not necessary for action or for science. It has no practical benefit. It is not psychologically compulsory. Loss of this belief does not constitute a skeptical problem worthy of the name. In fact, from the very beginning of the *Treatise* Hume signals his contempt for the search for ultimate principles, by which he means objective necessary connections (T Intro. 8–10). As for the unsatisfied desire to find these ultimate principles, it vanishes as soon as we see that the desired satisfaction is impossible (T Intro. 8–10). He never indicates that anything of importance is lost when we reject the idea of objective necessary connection. He presents his own alternative definitions of causation as fully adequate to the theoretical demands of common life and science (T 1.3.14.31).[33] Even in *Treatise* 1.4.7.5, he never explicitly states that belief

31. Garrett, *Hume*, 222. Cf. Garrett, *Cognition*, 221–22.

32. Thus also Ainslie, *Hume's True Scepticism*, 235; Henry Allison, "Hume's Philosophical Insouciance: A Reading of *Treatise* 1.4.7," *Hume Studies* 31 (2005), 335; Morris, "Hume's Conclusion," 95–96; Garrett, *Cognition*, 222; Singer, "Hume's Extreme Skepticism," 602. Some commentators do see a genuine skeptical threat to a core belief here though. Fogelin barely acknowledges T 1.4.7.5, but implies that it expresses Hume's serious concern "that enquiring into the operations of this faculty [imagination] has brought to light its arbitrary, weak, and capricious character." Fogelin, *Hume's Skeptical Crisis*, 128. Schmitt treats the problem as a real skeptical threat without ever questioning whether its target is a core belief. Schmitt, *Hume's Epistemology*, 348–54. Immanuel Kant takes Hume's criticism of the intelligibility of objective necessary connections to be his central skeptical argument. For Kant, causation without objective necessary connection is no longer the idea of causation at all, so Hume's alternative analysis of the causal relation is a nonstarter. "The very concept of a cause so obviously contains the concept of a necessity of connection with an effect and a strict universality of rule that it would be entirely lost if one sought, as Hume did, to derive it from a frequent association of that which happens with that which precedes and a habit (thus a merely subjective necessity) of connecting representations arising from that association." Kant, *Prolegomena to Any Future Metaphysics*, trans. Gary Hatfield, in *Theoretical Philosophy after 1781*, ed. Henry Allison and Peter Heath (New York: Cambridge University Press, 2002), B 5; 138. Hume and Kant may be able to agree on the fact that belief in causal relations is a core belief. But because they differ in their analysis of the concept of cause, Hume's argument constitutes a skeptical threat to the belief in causes by Kant's lights, but not by Hume's own lights. Kantian assumptions have perhaps led some interpreters to regard Hume's argument against objective necessary connections as one which Hume himself recognizes as a skeptical threat to a core belief. See for example De Pierris, "Hume's Pyrrhonian Skepticism," 363–64.

33. In the first *Enquiry*, Hume seems to imply that in losing an intelligible belief in objective necessary connections, we have lost something of importance, for which there is no fully satisfying replacement, even in his own two definitions of "cause" (EHU 7.28–30). However, despite his language of "scepticism" in Section 7, this material plays no role in the skeptical catalog of Section 12.

in objective necessary connections is a core belief, or that its loss presents a presents a problem for philosophy. He simply reiterates what he had already said in *Treatise* 1.3.14, that it comes as a shock when we realize this belief is an unintelligible falsehood.

Treatise 1.4.7.3 and 1.4.7.5 present skeptical problems for a wide swath of Hume's readers with assumptions that he himself does not share. *Treatise* 1.4.7.3 points out that if the vivifying principles of the unreasoning imagination do not defeasibly justify their products, then neither senses, memory, nor probable reasoning are justifying. *Treatise* 1.4.7.5 shows that if objective necessary connections are somehow indispensable for the justification of causal reasoning and science, then causal reasoning and science are unjustified. Hume does not explicitly affirm these assumptions himself. These assumptions conflict with his own stated principles of philosophy. The Title Principle does not seem to solve any of these skeptical problems.

The *reductio* arguments serve to motivate readers to adopt a purely practical response to epistemic skepticism. The Title Principle is just one form that a purely practical response might take; it is specifically suited to the defeater arguments from reason that Hume puts forward in *Treatise* 1.4.1–2. However, we can express Hume's response to skeptical challenges in a broader fashion. He advises us to practically moderate our commitment to our epistemic norms, so that we retain our core beliefs even when they fall short of our epistemic standards. Stated at this level of generality, his response applies to any sort of skeptical dilemma generated by any sort of epistemology against any sort of core belief. Is our (allegedly) core belief in objective necessary connections epistemologically unjustified because it is not produced by reason and violates the Copy Principle (T 1.4.7.5)? We should hold it anyway, on practical grounds. Are the beliefs produced by reason and the senses epistemologically defeated by rational arguments (as Hume sincerely holds, T 1.4.1–2)? We should hold them anyway, on practical grounds. We may disagree on which beliefs are core beliefs, or which epistemic standards they ought to meet, or whether they do meet those standards. But Hume's general point remains that for any given core belief, whatever the source of its negative epistemic status, we practically ought to retain the belief in spite of its epistemic status. This is so because giving up core beliefs is always disagreeable and harmful for ourselves and others.

Hume has good strategic reasons for choosing these particular *reductio* arguments. He is aware that his analyses of the causal connection and causal inference are the most radical and surprising contributions of Book 1 of the *Treatise*. Hume focuses on this material in the "Abstract" of the *Treatise*. On the other hand, his true skeptical problems (in T 1.4.1–2) are quite abstruse. He anticipates that his readers will sense a skeptical threat in his analysis of causation and inference in Part 3, but fail to appreciate the force of the defeater arguments he puts forward in Part 4 of Book 1. So in "Conclusion of this book" he recasts the material from Part 3 as if it really poses the skeptical threats that his readers sense, without discarding the skeptical problems that he himself faces. By one route or the other, he aims to drive his readers into skeptical dilemmas that will force them to moderate their practical commitment to their epistemic norms.

The reception of the *Treatise* seems to have convinced Hume of the difficulty and importance of making his skeptical challenges less abstruse and more widely accessible.[34] For he writes in the Preface to the "Abstract" of the *Treatise* that "The work, of which I here present the Reader with an abstract, has been complained of as obscure and difficult to be comprehended" (Preface 2). The apparent inability of his readers to understand the *Treatise* clearly motivates his attempt to explain it in the Abstract. Certainly, the *reductio* skeptical arguments he makes in paragraph 27 are far more accessible than those of *Treatise* 1.4.1–2.

The outcome of Hume's skeptical crisis is that our central beliefs are philosophically unjustified. But philosophy itself only has a limited practical and moral warrant. We are practically warranted in ignoring philosophy when it demands the suspension of core beliefs; we are practically warranted in carrying on with common life as well as research in the sciences. Hume's purely practical justification for holding core beliefs maintains a clear distinction between philosophical and practical normativity, yet gives him a good reason for continuing with his constructive scientific projects in the face of radical and irremediable philosophical skepticism. Hume's use of *reductio* arguments indicates that he does not stumble into epistemic skepticism by accident or in spite of his best efforts to avoid it. On the contrary, he drives his readers into a skeptical dilemma by any means available, even using arguments that he himself does not regard as compelling. This raises the question, "Why is Hume so intent on convincing us that some of our core beliefs are epistemologically unjustified?" The purely practical reading provides an excellent explanation: Hume uses epistemic skepticism to motivate readers to adopt a moderate practical attitude toward the epistemic demands of philosophy.

34. As I discuss further in chapter 6, Hume attributed the disappointing reception of the *Treatise* primarily to its form rather than to its matter.

Chapter Six

Philosophy and Justification in the *Enquiry*

This chapter explores the concept and norms of philosophy in the first *Enquiry*. Before examining the content of the first *Enquiry* in detail, I briefly compare its contents with those of the *Treatise* and examine Hume's remarks about the two works. This evidence indicates they share the same basic philosophical principles, but express and apply them in different ways. Turning to the *Enquiry*, I argue that, as in the *Treatise*, Hume uses the term "philosophy" to refer to a normative method of inquiry and belief-formation that governs all of the special sciences. The *Enquiry* has no passage equivalent to *Treatise* 1.4.4.1–2 that draws a broad normative distinction between PIU and CWI principles of the imagination. The *Enquiry* emphasizes instead that philosophy stays within the limits of the human understanding. It identifies many familiar philosophically approved doxastic processes and practices, including intuition and demonstrative reasoning, the formation of sensory beliefs, memory, and experience-based causal inference. One of the major goals of the first *Enquiry* is to demarcate the limits of human understanding and thereby expose pseudo-philosophy for what it is. As in the *Treatise*, philosophy aims at truth, not at practical benefits. Although it has real drawbacks, philosophy has enough practical benefits to justify a moderate pursuit of it by those so inclined.

THE RELATIONSHIP BETWEEN THE *TREATISE* AND THE *ENQUIRY*

Hume recast Book 1 of the *Treatise* in *Philosophical Essays concerning Human Understanding* (1748), later retitled *An Enquiry Concerning Human Understanding* (1756). The *Enquiry* drops the more arcane aspects of Book 1 of the *Treatise*, focuses on Hume's theory of causation and causal inference, and adds some (mainly) anti-religious applications of that theory. The result is a far more streamlined and readable work. As Hume wrote to Gilbert Elliot of Minto in 1751, "By shortening and simplifying the Questions, I really render them much more complete. *Addo dum minuo*" (HL i.158).

Large parts of the *Enquiry* correspond closely to parts of Book 1 of the *Treatise*. Section 1, "Of the Different Species of Philosophy," like the "Introduction" to the *Treatise*, defends the value and method of undertaking a scientific investigation of the human mind, though from a fresh angle, as I discuss later in the chapter. Section 2, "Of the Origin of Ideas," sets forth in a few pages Hume's theory of perceptions (cf. T 1.1.1–3) culminating in the Copy Principle. Section 3, "Of the Association of Ideas," discusses the three associative principles (cf. T 1.1.4), adding some illustrations from literature. Hume omits from the *Enquiry* the *Treatise* material on philosophical relations, modes and substances, and abstract ideas (T 1.1.1–5), as well as his theory of space and time (T 1.2). Section 4, "Sceptical Doubts concerning the Operations of the Understanding," presents the negative side of his theory of causal inference (cf. T 1.3.1–6), showing that causal inference is not produced by reason. Section 5, "Sceptical Solution of these Doubts," presents the positive side of his theory of causal inference (cf. T 1.3.7–10), showing how custom produces belief. Section 6, "Of Probability," summarizes *Treatise* 1.3.11–12. The material from *Treatise* 1.3.13, "Of Unphilosophical Probability," seems to drop out of the *Enquiry*. So too do the "Rules by which to judge of causes and effects" (T 1.3.15). Section 7, "Of the Idea of Necessary Connexion," corresponds to the section the same name in the *Treatise* (T 1.3.14). Section 9, "Of the Reason of Animals," corresponds to the *Treatise* section of the same name (T 1.3.16).

Three sections of the *Enquiry* have no parallel in Book 1 of the *Treatise*. Section 8, "Of Liberty and Necessity," applies Hume's concept of causation to matters of the will. In Part 1, Hume defends a form of compatibilism (cf. T 2.3.1–3, whose first section is entitled "Of liberty and necessity"). Part 2 argues that if God is the first cause of all human actions, then he bears responsibility for whatever turpitude they possess. Sections 10–11 do not correspond to anything in any book of the *Treatise*. Both sections deploy Hume's theory of causal inference to criticize commonplace arguments of religious apologists. Section 10, "Of Miracles," argues that no testimonial evidence can ever justify belief in the occurrence of a miracle. Section 11, "Of a Particular Providence and of a Future State," argues that experiential evidence does not support belief in a morally perfect deity who will eventually dole out more justice than we currently observe in the world.

Section 12, "Of the Academical or Sceptical Philosophy," corresponds broadly to Part 4 of Book 1 of the *Treatise*. In Part 1, Hume puts forward skeptical arguments against the senses—specifically against the vulgar belief in body, the philosophical system of double existences, and the modern philosophy of primary and secondary qualities (cf. T 1.4.2, 1.4.4). However, the arguments against substances composed of accidents inhering in a substratum (cf. T 1.4.3, 1.4.5, 1.4.6) drop out of the *Enquiry*. Part 2 advances skeptical arguments against demonstrative and probable reasoning. Although this topic corresponds to the topic of *Treatise* 1.4.1, the argument for the iterative self-defeat of reasoning does not appear. Instead, Hume uses material from his theory of spatial and temporal quantity (cf. T 1.2) to generate a skeptical argument against abstract reasoning, and he uses his account of causal inference (cf. T 1.3.6, EHU 4–5) to suggest a skeptical argument against inductive reasoning. In the latter part of Part 2 and in Part 3, Hume responds to these skeptical worries and charts a course forward (cf. T 1.4.7).

Hume's remarks about the *Treatise* and the *Enquiry* indicate that their differences have to do primarily with manner, though also to a lesser extent with matter.[1] His primary stated disappointment with the *Treatise* seems to have been that it did not attract the attention he had hoped it would.

> Never literary attempt was more unfortunate than my Treatise of Human Nature. It fell *dead-born from the press*, without reaching such distinction, as even to excite a murmur among the zealots. (MOL xxxiv)

He attributed this lack of success primarily to the style and manner of the *Treatise* rather than to its content, and thus sought to recast the content in a more appealing form in the *Enquiry*.

> I had always entertained a notion, that my want of success in publishing the Treatise of Human Nature had proceeded more from the manner than the matter, and that I had been guilty of a very usual indiscretion, in going to the press too early. I, therefore, cast the first part of that work anew in the Enquiry concerning Human Understanding. (MOL xxxv)

Hume does not say that there were no problems with the "matter" of the *Treatise*, but he puts more weight on its manner. The "Advertisement" at the front of the 1777 edition of *Essays and Treatises on Several Subjects* (containing the first *Enquiry*) gives a similar impression. Hume writes that the author "cast the whole [*Treatise*] anew in the following pieces, where some

1. I am indebted in this paragraph to Qu's thorough discussion of all of Hume's relevant remarks. Qu, *Hume's Epistemological Evolution*, 8–13. Qu comes to the conclusion that these remarks do not settle the case for or against philosophical continuity.

negligences in his former reasoning and more in the expression, are, he hopes, corrected." The *Enquiry* does make substantive corrections to the arguments of the *Treatise*—that is, to "some negligences in his former reasoning." But put this way, the "negligences" sound like matters of detail rather than shifts in major principles. The *Enquiry* makes more corrections to "the expression" of the *Treatise*. Hume writes to Elliot that "The philosophical Principles are the same in both" (HL i.158). In a 1754 letter to John Stewart, Hume writes that the *Enquiry* contains "the same Doctrines" as the *Treatise*, though "better illustrated and exprest" (HL i. 187). Hume did place great importance on the differences between the two works. He concludes the "Advertisement" by stating that "Henceforth, the Author desires, that the following Pieces may alone be regarded as containing his philosophical sentiments and principles"—in effect disavowing the *Treatise* in favor of the *Enquiry* and its companion works. But his statements indicate that he disavowed the *Treatise* primarily because of its manner of expression; he saw its philosophical content as sound, aside perhaps from a "some negligences."

One negligence that the *Enquiry* amends is Hume's articulation of his skeptical stance. In *Letter to a Gentleman*, Hume complains that the earliest critic of the *Treatise* grossly misconstrued him as advocating a total suspension of all beliefs. He attempts to explain the true meaning of the *Treatise* in new terms—namely, by distancing himself from what he now calls Pyrrhonian skepticism. He similarly disavows Pyrrhonian skepticism in Section 12 of the first *Enquiry*. So one of the key changes in the manner of the first *Enquiry* is his way of explaining the skeptical stance he had taken up already in the *Treatise*. The matter of his skepticism had remained the same, though he thought he had improved his manner of expressing it.

Despite his clarificatory efforts in the *Enquiry*, critics such as James Beattie and Thomas Reid continued to accuse Hume of adopting an unlivable excessive skepticism. In a letter to his publisher, William Strahan, Hume wrote of the 1777 Advertisement that "It is a compleat Answer to Dr Reid and to that bigoted silly Fellow, Beattie" (HL ii.301). The "Advertisement" is not itself an answer to Reid and Beattie; it only points to the works in *Essays and Treatises*, including the first *Enquiry*, as containing an answer to them. But the first *Enquiry* itself long antedated the criticisms of Beattie and Reid. It "answers" them only by stating, even more clearly than the *Treatise*, that Hume is not a Pyrrhonian skeptic who advocates the suspension of all beliefs.[2]

2. Cf. Harris, *Hume*, 443. By contrast, Qu says that Hume answers Reid and Beattie "by pointing out the inadequacies of their epistemological position, while on his part providing more philosophically tenable arguments against excessive scepticism." Qu, *Hume's Epistemological Evolution*, 21. At most, the *Enquiry* might presciently point out the inadequacies of a position similar to Reid's or Beattie's, since their actual positions had not yet appeared.

All told, I think the burden of proof lies on those who seek to show discontinuity between the philosophical doctrines of the two works. It counts in favor of my reading that it ascribes a large degree of material continuity to them. However, I do not simply take continuity for granted on any particular issue. In this chapter, I try to examine the *Enquiry* entirely on its own terms, not presuming continuity with the *Treatise* but noting when it emerges.

THE CONCEPT OF PHILOSOPHY

Hume introduces the leading motifs of his concept of philosophy immediately in Section 1, "Of the Different Species of Philosophy." The section begins with a discussion of two species of philosophers. The first species popularizes the science of human nature for the moral edification of a broad reading public (EHU 1.1). Philosophers of the second species make the scientific discoveries and write primarily for their scholarly peers (EHU 1.2). This second species of philosophy is philosophy proper, the normative practice of scientific inquiry, theory construction, and belief formation. Hume's main point in Section 1 is to give a defense of the practice of the second species of philosophy. He discloses much about the nature of philosophy in the course of this defense.

From the first sentence of Section 1, Hume uses "philosophy" virtually interchangeably with "science." He can speak of both in general, unqualified terms. For example, speaking of philosophers, he says that they "rest not satisfied till they arrive at those original principles, by which, in every science, all human curiosity must be bounded" (EHU 1.2). The branches of philosophy are, by that very fact, branches of science. "Moral philosophy," for example, is just another name for "the science of human nature" (EHU 1.1). Hume also speaks of philosophy almost interchangeably with reason, reasoning, and metaphysics. The "abstract and profound" species of philosophy consists in "profound reasonings, or what is commonly called *metaphysics*" (EHU 1.7; cf. 1.11). Metaphysics is a science, albeit a contested one, and in fact "the most contentious science" (EHU 8.23). Hume's defense of profound philosophy is at the same time a defense of "true metaphysics" (EHU 1.12). Later, Hume speaks of "moral philosophy" interchangeably with "moral or metaphysical sciences," and contrasts them with "natural philosophy" and "mathematics" (or "mathematical sciences") (EHU 7.2, 1).

Philosophy is essentially characterized by painstaking adherence to reason in pursuit of accuracy and truth. Though their task is "arduous," philosophers "are deterred by no difficulties"; they "think themselves sufficiently compensated for the labour of their whole lives, if they can discover some hidden truths, which may contribute to the instruction of posterity" (EHU 1.2). Their speculations are "abstract and profound" (EHU 1.7; cf. 1.16), "abstruse" (EHU 1.3), "unintelligible to common readers" (EHU 1.2), "of difficult comprehension" (EHU 1.16). But the abstract reasoning of philosophy is also "accurate" (EHU 1.3, 1.8, 1.12, 1.13). "Probity and accuracy . . . are the natural result of a just philosophy" (EHU 1.6). The "genius of philosophy" diffuses a "spirit of accuracy" in "every art and calling" (EHU 1.9).

Hume distinguishes between philosophy de jure and philosophy de facto in the *Enquiry*, just as he does in the *Treatise*. The chief difference between good and bad philosophy lies in the observance of the proper scope limitation on philosophical enquiry. Hume shares the concern that, in practice, much actual philosophy violates a normative scope restriction by speculating on topics that lie beyond the reach of the understanding. But he thinks that the most effective antidote is not to give up on all philosophy. Instead, we must engage in profound moral philosophy in order to clearly demarcate, once and for all, the capacities and limits of the understanding. "We . . . must cultivate true metaphysics with some care, in order to destroy the false and adulterate" (EHU 1.12). Only a philosophy characterized by "accurate and just reasoning," can conquer "this deceitful philosophy . . . that abstruse philosophy and metaphysical jargon" (EHU 1.12). The deceitful philosophy has only "the air of science and wisdom," but not the reality (EHU 1.12).

THE NORMS OF PHILOSOPHY AND THE LIMITS OF HUMAN UNDERSTANDING

The limits of the human understanding constitute the normative limits of legitimate philosophical inquiry. Hume indicates this in passing at different points. Science provides relatively "little satisfaction" because "the bounds of human understanding" are so narrow (EHU 1.6). Hume argues that his own project of mental geography does not reach "beyond the compass of human understanding" (EHU 1.14). Regrettably, many philosophers (especially the superstitious philosophical theologians) transgress this limit.

> Here indeed lies the justest and most plausible objection against a considerable part of metaphysics, that they are not properly a science; but arise . . . from the fruitless efforts of human vanity, which would penetrate into subjects utterly inaccessible to the understanding. (EHU 1.11)

Hume proposes to combat this kind of pseudo-philosophy by marking out more precisely the limits of the understanding:

> The only method of freeing learning, at once, from these abstruse questions, is to enquire seriously into the nature of human understanding, and shew, from an exact analysis of its powers and capacity, that it is by no means fitted for such remote and abstruse subjects. (EHU 1.12)

At the very end of the *Enquiry*, he reports the results of his investigation, and lists "what are the proper subjects of science and enquiry" (EHU 12.25). The "only objects of the abstract sciences or of demonstration are quantity and number" (EHU 12.27). Any other factual claim can only be established by probable reasoning based on experience (EHU 12.32). "Morals and criticism are not so properly objects of the understanding as of taste and sentiment," though reason can have a supporting role in these subjects (EHU 12.33).[3] In the last sentence of the book, he famously recommends what we should do with a volume of "divinity or school metaphysics" that does not contain such abstract or experimental reasoning: "Commit it then to the flames: For it can contain nothing but sophistry and illusion" (EHU 12.34). So his demarcation of the limits of human understanding does free us, in the end, from false metaphysics, as promised in Section 1.

Demonstrative Reasoning, Intuition, and Clear Definition

Philosophy endorses intuition and demonstration, the mental operations whereby we discover relations of ideas. Reason has two kinds of objects: relations of ideas and matters of fact.

> ALL the objects of human reason or enquiry may naturally be divided into two kinds, to wit, *Relations of Ideas* and *Matters of Fact*. Of the first kind are the sciences of Geometry, Algebra, and Arithmetic; and in short, every affirmation, which is either intuitively or demonstratively certain. (EHU 4.1)

We know relations of ideas either by way of intuition or demonstration (EHU 12.27). The only thing that interferes with our intuitive pronouncements is definitional vagueness. For example, "to convince us of this proposition, *that where there is no property, there can be no injustice*, it is only necessary to define the terms, and explain injustice to be a violation of property. This proposition is, indeed, nothing but a more imperfect definition" (EHU 12.27). Hume has a favorite tool for clarifying definitions: the Copy Principle.

> When we entertain, therefore, any suspicion, that a philosophical term is employed without any meaning or idea (as is but too frequent), we need but enquire, *from what impression is that supposed idea derived?* (EHU 2.9)

3. By "criticism," Hume means claims about aesthetics.

Once our ideas have been clarified using the Copy Principle, we can make intuitively certain judgments about claims that are true (or false) by definition. Philosophy requires definitions which are as precise and exact as possible—but no more so. For example, Hume acknowledges that his definition of belief "may seem . . . unphilosophical," but explains that "in philosophy, we can go no farther" than to describe the distinctive feeling in the mind (EHU 5.12; cf. T 1.3.7.7).

Hume expresses confidence in the ability of philosophers to resolve questions and reach consensus, if they simply base their factual claims on experience and clearly define their terms. Reflecting on the long dispute over liberty and necessity, he remarks

> But if the question regard any subject of common life and experience; nothing, one would think, could preserve the dispute so long undecided, but some ambiguous expressions, which keep the antagonists still at a distance, and hinder them from grappling with each other. (EHU 8.1)

Experience-Based Causal Inference

Most of the *Enquiry* makes it unambiguously clear that Hume regards inductive reasoning—that is, experience-based causal inference—as a cognitive process that philosophically justifies its conclusions. His famous statement in the section on miracles epitomizes this attitude:

> A wise man, therefore, proportions his belief to the evidence. In such conclusions as are founded on an infallible experience, he expects the event with the last degree of assurance, and regards his past experience as a full *proof* of the future existence of that event. In other cases, he proceeds with more caution: He weighs the opposite experiments: He considers which side is supported by the greater number of experiments: To that side he inclines, with doubt and hesitation; and when at last he fixes his judgment, the evidence exceeds not what we properly call *probability*. (EHU 10.4)

Hume's objection to the acceptance of miracle testimony and to arguments for a particular providence is that they violate the norms of inductive reasoning. On the other hand, all of his positive conclusions about human nature are based on inductive reasoning.

However, some passages understandably raise doubt about whether Hume regards causal inferences as philosophically justified after all. The titles of Section 4 ("Sceptical Doubts concerning the Operations of the Understanding) and Section 5 ("Sceptical Solution of these Doubts") create the impression that causal inference faces a skeptical problem, and that the only "solution" on hand fails to dispel it. The main conclusions of these sections are familiar from enough from Book 1, Part 3 of the *Treatise*.[4] On one hand, causal inferences are not produced by reason: "even after we have experience of the operations of cause and effect, our conclusions from that experience are *not* founded on reasoning, or any process of the understanding" (EHU 4.15). On the other hand, causal inferences are produced by the principle of "custom or habit" (EHU 5.5). Later, Hume describes these same findings as forceful philosophical objections of the skeptic (EHU 12.22).

I address the alleged skeptical force of Hume's account of causal inference in the next chapter. If there are any skeptical implications of Sections 4 and 5—if the claims in these sections do show that causal inferences are unjustified—then Hume brackets them out of the rest of the book. Outside of Sections 4, 5, and parts of 12, he everywhere implicitly and explicitly treats proper inductive reasoning as philosophically normative. It is only when speculations go beyond the bounds of experience-based reasoning that they transgress the scope limitation of philosophy.

Hume's investigations show that the understanding has the power to make causal inferences only on the basis of experience. This is the major lesson of Part 1 of Section 4. "All reasonings concerning matter of fact seem to be founded on the relation of *Cause* and *Effect*" (EHU 4.4), and Hume lays it down "as a general proposition, which admits of no exception, that the knowledge of this relation . . . arises entirely from experience, when we find, that any particular objects are constantly conjoined with each other" (EHU 4.6). So any belief we hold about a matter of fact that lies "beyond the evidence of our memory and senses" (EHU 4.4) depends upon experience. Of course, humans frequently make inferences that lack any grounds in experience. But these are bad arguments, inferences that do not justify their conclusions. In these cases, the understanding either malfunctions or suffers interference from the passions or other untoward influences. The scope limitation of philosophy limits inductive inferences to those based on experience.

Hume does not limit the subjects about which we can reason from experience, although he often sounds like he does. He warns against any attempt to "penetrate into subjects utterly inaccessible to the understanding" (EHU 1.11). The understanding is "is by no means fitted for such remote and abstruse subjects" (EHU 1.12). At times it sounds as though Hume would proscribe otherwise irreproachable reasoning simply because it concerns

4. Hume's arguments for these conclusions do differ in points of detail, especially in mode of expression. See especially Millican, "Hume's 'Scepticism' About Induction"; Qu, *Hume's Epistemological Evolution*, chapters 3–4.

subjects that are too high and distant. Consider for example his criticism of the doctrine of occasionalism:

> It seems to me, that this theory of the universal energy and operation of the Supreme Being, is too bold ever to carry conviction with it to a man, sufficiently apprized of the weakness of human reason, and the narrow limits, to which it is confined in all its operations. Though the chain of arguments, which conduct to it, were ever so logical, there must arise a strong suspicion, if not an absolute assurance, that it has carried us quite beyond the reach of our faculties, when it leads to conclusions so extraordinary, and so remote from common life and experience. We are got into fairy land, long ere we have reached the last steps of our theory; and *there* we have no reason to trust our common methods of argument, or to think that our usual analogies and probabilities have any authority. Our line is too short to fathom such immense abysses. And however we may flatter ourselves, that we are guided, in every step which we take, by a kind of verisimilitude and experience; we may be assured that this fancied experience has no authority, when we thus apply it to subjects, that lie entirely out of the sphere of experience. But on this we shall have occasion to touch afterwards [footnote 15: Section 12]. (EHU 7.24)

Hume would reject occasionalism "Though the chain of arguments, which conduct to it, were ever so logical," simply because that chain of arguments "leads to conclusions so extraordinary, and so remote from common life. We are got into fairy land, long ere we have reached the last steps of our theory; and *there* we have no reason to trust our common methods of argument, or to think that our usual analogies and probabilities have any authority." He sounds as though he is claiming that otherwise irreproachable arguments that reach conclusions on matters beyond the banal concerns of common life are simply not reliable or trustworthy. He gives the scope restriction an air of piety when he adds "Our line is too short to fathom such immense abysses," echoing theological commonplaces about divine transcendence.

But a second look at the passage makes it clear that Hume rejects occasionalist reasoning because it is not based on real experience. We merely "flatter ourselves, that we are guided, in every step which we take, by a kind of verisimilitude and experience," but in fact it is only "fancied experience" that "has no authority." The occasionalist's argument reaches "to subjects, that lie entirely out of the sphere of experience" (EHU 7.24). (As Philo says, "We have no experience of divine attributes and operations" [DNR 2.4]). So the subjects that Hume proscribes are only those "that lie entirely out of the sphere of experience."

If Hume intends to forbid otherwise irreproachable reasoning simply because of its subject matter, then he would refuse to discuss arguments about the divine existence and operations. If reasoning about the divine was unreliable in principle, regardless of the merits of the arguments or the conclusions reached, then such reasoning would be pointless. But he does engage these arguments throughout his works, notably in *Enquiry* Section 11 and the *Dialogues*. In his philosophy of religion, he does not criticize his opponents for making arguments, but for making bad arguments. He consistently points out arguments not adequately grounded in experience. This is exactly what we should expect if the scope restriction fundamentally concerns the basis of arguments rather than their subject matter.

For example, in Section 11, the main character discusses natural theology but criticizes arguments for having inadequate grounding in experience. The narrator's skeptical friend insists that "A philosopher" cannot approve of "those vain reasoners" who derive their ideas of the gods from anything other than "the present phaenomena" (EHU 11.21). He ends his harangue with the following ringing declamation:

> The experienced train of events is the great standard, by which we all regulate our conduct. Nothing else can be appealed to in the field, or in the senate. Nothing else ought ever to be heard of in the school, or in the closet. In vain would our limited understanding break through those boundaries, which are too narrow for our fond imagination. (EHU 11.23)

The first-person narrator specifically approves of the interlocutor's view: "But allowing you to make experience (as indeed I think you ought) the only standard of our judgment concerning this" (EHU 11.24). The natural theology arguments under consideration fall woefully short of this standard.

Experience, Common Life, and Philosophy

Hume limits philosophy and reason to the sphere of experience, but he also sometimes states this scope limitation in terms of common life. Consider the following texts. In the passage about occasionalism quoted above, he rejects reasoning that "leads us to conclusions so extraordinary, and so remote from common life and experience" (EHU 7.24). The proper sphere of philosophy lies within "the reach of human capacity" and the subjects "of common life and experience" (EHU 8.1). "The power of philosophy" reaches only as far as "natural and unassisted reason," and "her true and proper province" is "the examination of common life" (EHU 8.36). "All the philosophy, therefore, in the world . . . will never be able to carry us beyond the usual course of experience, or give us measures of conduct and behavior different from those which are furnished by reflections on common life" (EHU 11.27). Mitigated skeptics recognize that "Philosophical decisions are nothing but the reflections of common life, methodized and corrected," and they are never "tempted to go beyond common life" (EHU 12.25).[5] These passages lead us to think that philosophy and common life are coextensive with one another and with the sphere of experience-based reasoning.

On the other hand, Hume also sometimes seems to distinguish between common life and philosophy. In several texts he points out that common life and philosophy agree in some respect—which suggests not only that the two are distinct, but that they often differ. Consider the following examples. Hume says that "*belief* . . . is a term, that every one sufficiently understands in common life. And in philosophy, we can go no farther than assert, that *belief* is something felt by the mind, which distinguishes the ideas of the judgment from the fictions of the imagination" (EHU 5.12). Since we have no impression of a "connexion or power" between objects, "these words are absolutely without any meaning, when employed either in philosophical reasonings, or common life" (EHU 7.26). The regular conjunction between motives and voluntary actions "has never been the subject of dispute, either in philosophy or common life" (EHU 8.16; cf. 8.27). "The maxims of common life are subjected to the same doubt as the most profound principles or conclusions of metaphysics and theology," by a certain species of philosophers (EHU 12.5). All of these texts treat the coincidence of common life and philosophy as noteworthy and exceptional. But if philosophy and common life are identical or at least necessarily coextensive, then it should go without saying that they always coincide.

Other texts suggest a positive divergence between the common life and philosophy. At the very outset of the *Enquiry*, Hume seems to depict common life and philosophy as mutually exclusive. The first species of philosophy "enters more into common life," but the second species (philosophy proper) does not:

> On the contrary, the abstruse philosophy, being founded on a turn of mind, which cannot enter into business and action, vanishes when the philosopher leaves the shade, and comes into open day; nor can its principles easily retain any influence over our conduct and behaviour. The feelings of our heart, the agitation of our passions, the vehemence of our affections, dissipate all its conclusions, and reduce the profound philosopher to a mere plebian. (EHU 1.3)

The last section of the *Enquiry* sounds similar notes. The ideas of space and time are "ideas, which, in common life and to a careless view, are very clear and intelligible, but when they pass through the scrutiny of the profound sci-

5. Note that confining philosophy to common life is not unique to mitigated skepticism. The mitigated skeptic simply has further motivation to appreciate the scope limitation of philosophy *simpliciter*.

ences . . . afford principles, which seem full of absurdity and contradiction" (EHU 12.18). Pyrrhonism "may flourish and triumph in the schools," but is subverted by "the occupations of common life," because the schools are the "proper sphere" for the sceptic's "*philosophical* objections" (EHU 12.21–22).

Hume's idea seems to be that common life and philosophy both involve reasoning from experience, but we reason farther in the latter than we do in the former. Since we reason from experience in both common life and philosophy, the two are coextensive in an important sense. But we make only rudimentary inferences in common life. Although we constantly make simple causal inferences, we lack the ability to track arguments (good or bad) of much depth and abstraction while actively engaged in quotidian physical tasks. Philosophy consists in carrying our experience-based reasoning farther and deeper than we can in common life. This requires the undistracted, disciplined intellectual pursuit that Hume calls "philosophy." In this sense, philosophy includes common life but extends beyond it; we might even say that philosophy, properly speaking, begins where the rudimentary inferences of common life end.

There is an important difference between arguments that reach beyond the realm of common life and experience, and arguments that lose their grounding in common life and experience. Good reasoning begins with experience as its basis and reaches conclusions beyond the sphere of common life. It is no objection to the second species of philosophers that "their speculations seem abstract, and even unintelligible to common readers" (EHU 1.2). But false reasoning reaches conclusions that are not grounded in common life, the sphere of experience. This is why Hume objects that occasionalism is "so remote from common life and experience" (EHU 7.24).

Hume also depicts philosophy as an extension of the same reasoning of common life in *Dialogues* 1.9. When a man "philosophizes, either on natural or moral subjects," he "carries his speculations farther than . . . necessity constrains him"—that is, beyond the boundaries of common life. In this sense, philosophy begins only where common life ends.[6] But if by philosophy we

6. Compare *Treatise* 1.4.7.13, where Hume says that we need philosophy as our guide only when we leave the sphere of common life: "Since therefore 'tis almost impossible for the mind of man to rest, like those of beasts, in that narrow circle of objects, which are the subject of daily conversation and action, we ought only to deliberate concerning the choice of our guide, and ought to prefer that which is safest and most agreeable. And in this respect I make bold to recommend philosophy, and shall not scruple to give it the preference to superstition of every kind or denomination." Qu argues that Hume changed his mind between the *Treatise* and the *Enquiry* about whether philosophy falls within the sphere of common life. Qu, *Hume's Epistemic Evolution*, 99–100, 205n42. However, *Treatise* 1.4.7.13 need not reflect a substantively different position than in the *Enquiry* and *Dialogues*. In the later works, Hume can speak of philosophy in different senses as either confined to common life or beginning where common life ends.

mean simply "reasoning from experience," then philosophy begins already within common life:

> everyone even in common life, is constrained to have more or less of this philosophy; . . . from our earliest infancy we make continual advances in forming more general principles of conduct and reasoning; . . . the larger experience we acquire, and the stronger reason we are endowed with, we always render our principles the more general and comprehensive; and . . . what we call *philosophy* is nothing but a more regular and methodical operation of the same kind. To philosophize on such subjects is nothing essentially different from reasoning on common life; and we may only expect greater stability, if not greater truth, from our philosophy, on account of its exacter and more scrupulous method of proceeding. (D 1.9)

The last sentence signals the difference between common life and philosophy: philosophy has "exacter and more scrupulous method of proceeding." The philosopher reasons more carefully, but not in an essentially different way, than we do in common life.

Senses and Memory

Sensory and memorial beliefs are also treated as philosophically justified in the *Enquiry*. The justificatory authority of experience entails the justificatory authority of senses and memory. Experience consists of sense perceptions, observations, often retained in memory. If the observations or memories of observations on which we base our inferences are false, then our conclusion is not justified. Hume makes this point when he rejects the arguments for occasionalism because they are based only on "fancied experience" (EHU 7.24). Another indication of the justificatory authority of sense beliefs and memory is that they can defeat the justification of a conclusion of probable reason, as I discuss below.

Belief in the senses means, for Hume, belief in the existence and qualities of mind-independent objects that correspond to our sense perceptions.[7] Just in passing, Hume mentions "external bodies, the objects of our senses" (EHU 1.13). More importantly, when he takes up skepticism about the senses in Section 12, the skeptical arguments all target different kinds of belief in mind-independent objects. All humans and animals believe that their senses

7. I make this point in opposition to the suggestion of Millican and Qu that Hume could countenance sensory beliefs that were not beliefs about the qualities of mind-independent objects at all. Millican, "Hume's Chief Argument," 99, 100; Qu, *Hume's Epistemological Evolution*, 192–93. I do not think Hume would find this suggestion intelligible. I discuss this point further in the next chapter.

inform them of a mind-independent world (EHU 12.7). Our first instinct is to "suppose the very images, presented by the senses, to be the external objects" (EHU 12.8). A bit of philosophy teaches us that the perceptions present to our mind are merely "fleeting copies or representations of other existences, which remain uniform and independent" (EHU 12.9). "Modern enquirers" hold that only the primary qualities of those perceptions present to our minds represent the qualities of mind-independent objects (EHU 12.15). If we give up belief in all of the particular qualities of mind-independent objects, we have surrendered to the skeptic:

> Bereave matter of all its intelligible qualities, both primary and secondary, you in a manner annihilate it, and leave only a certain unknown, inexplicable *something*, as the cause of our perceptions; a notion so imperfect, that no sceptic will think it worth while to contend against it. (EHU 12.16)

In other words, sensory belief entails more than simply belief that mind-independent objects exist. Sensory belief involves belief in the qualities of these objects, knowledge of which qualities our sense perceptions convey to us. To give up belief that the senses convey knowledge of the qualities of mind-independent objects is to give up the only kind of sensory belief worth arguing about. The class of sensory beliefs in the existence and qualities of mind-independent objects is a core class of beliefs. This is a point that is crucial to the evaluation of the force of Hume's skeptical arguments in Section 12 of the *Enquiry*, which I will take up in the next chapter.

Epistemic Defeat

Hume's discussion of testimony for miracles in Section 10 implies certain broad principles of justificatory defeat. The first paragraph lays down a principle for adjudicating between inductive and sensory evidence. "A weaker evidence can never destroy a stronger"; to assent to a proposition contradicted by stronger evidence is "directly contrary to the rules of just reasoning" (EHU 10.1). Since all inductive arguments must be based on sensory evidence, sensory evidence is never weaker than the evidence of an inductive argument. Therefore, it is "contrary to the rules of just reasoning" to assent to an inductive inference whose conclusion "contradicts sense" (EHU 10.1).

Tillotson's rejection of the doctrine of the real presence provides one example of the appropriate deployment of this defeater structure (EHU 10.1). The Indian prince's disbelief in the existence of ice provides a similar example. On one hand, he had some testimonial (inductive) evidence in support of the freezing effects of cold weather. On the other hand, he had stronger evidence against it, from disanalogy to his own experience.

> The INDIAN prince, who refused to believe the first relations concerning the effects of frost, reasoned justly; and it naturally required very strong testimony to engage his assent to facts, that arose from a state of nature, with which he was unacquainted, and which bore so little analogy to those events, of which he had had constant and uniform experience. Though they were not contrary to his experience, they were not conformable to it [22]. (EHU 10.10)

In this case, a higher degree of probabilistic evidence on one side outweighs a lesser degree of probabilistic evidence on the other. Causal arguments (broadly construed to include arguments from analogy and from testimony) lie on either side of the question. Hume sums up this defeater principle later when he says

> It is experience only, which gives authority to human testimony; and it is the same experience, which assures us of the laws of nature. When, therefore, these two kinds of experience are contrary, we have nothing to do but subtract the one from the other, and embrace an opinion, either on one side or the other, with that assurance which arises from the remainder. (EHU 10.35)

All inductive evidence depends on experience, so equal degrees of experimental evidence cancel one another out.

Moderate Antecedent Skepticism; Miscellaneous Norms

A moderate version of what Hume calls "antecedent" skepticism is "a necessary preparative to the study of philosophy" (EHU 12.4).[8] Moderate antecedent skepticism cultivates intellectual traits which are normative for philosophy: "impartiality in our judgments" and freedom "from all those prejudices, which we may have imbibed from education or rash opinion" (ibid.). It disposes us to follow the normative method of philosophy, which is "To begin with clear and self-evident principles, to advance by timorous and sure steps, to review frequently our conclusions, and examine accurately all their consequences" (ibid.). This method is necessary for the attainment of whatever knowledge is within the reach of the human intellect:

> Though by these means we shall make both a slow and a short progress in our systems; are the only methods, by which we can ever hope to reach truth, and attain a proper stability and certainty in our determinations. (Ibid.)

8. Note that three out of the four forms of skepticism Hume discusses in Section 12 have distinctive names: Cartesian (excessive antecedent skepticism), Pyrrhonian (excessive consequent skepticism), and Academic (moderate consequent skepticism). But moderate antecedent skepticism has no name. It is simply constitutive of philosophy *simpliciter*.

Philosophical inquiry ought "To begin with clear and self-evident principles" (EHU 12.4). There are many such principles, as Hume indicates in his critique of Descartes. Cartesian skepticism insists that we prove the veracity of our faculties "by a chain of reasoning, deduced from some original principle, which cannot possibly be fallacious or deceitful" (EHU 12.3). Hume rejoins that there is no "such original principle, which has a prerogative above others, that are self-evident and convincing" (EHU 12.2). In other words, there are many "self-evident and convincing" principles. (He does not say that these principles "cannot possibly be fallacious or deceitful," as Descartes wishes). From these foundational premises, philosophy ought "to advance by timorous and sure steps" (EHU 12.4).

Hsueh Qu, following the lead of Garrett and Millican, argues that "Hume accords default or prima facie authority to our faculties"—that is, to all of our faculties, not merely the subset of our faculties of which philosophy approves.[9] He reasons that since Hume has no a priori epistemic principles, distinctions between faculties must arise a posteriori.[10] There must be a process of winnowing before philosophy can rule out, say, the propensity to project our emotions onto external objects.

But I think that this reaches beyond what Hume actually has in view in this passage. Qu may be right that Hume ought to initially grant default authority to all faculties. Perhaps Hume's view does require a process of winnowing to occur between birth and the commencement of philosophy proper, a process in which untoward faculties evoke disapproval. Be that as it may, in *Enquiry* 12.3, Hume has in view only the result or precipitate of that winnowing process. He has in view the moment when we are about to embark on philosophy, understood in the normative way he consistently describes it in his works. He pictures us at the starting line, as it were, of philosophical investigations. Antecedent to philosophy, at the starting line, some but not all faculties possess default authority. However the winnowing happens—if it indeed that is what happens—it is already done. Reason, the senses, and so forth have authority, but (for example) the propensity to project does not.

Hume makes several other general, isolated remarks about the normative conditions or requirements of philosophy throughout the *Enquiry*. His second philosophical objection to occasionalism invokes the principle of parity of reasoning (EHU 7.25).[11] "Philosophy . . . requires entire liberty above all other privileges, and chiefly flourishes from the free opposition of sentiments

9. Qu, *Hume's Epistemic Evolution*, 183.

10. Ibid., 183n7.

11. His line of thought seems to run as follows: we are as ignorant of the causal powers of minds (including the divine mind) as of bodies. If ignorance of some fact justifies denial of it, then we ought to deny the causal power of both minds and bodies. If ignorance does not justify denial, then we ought to deny the causal powers of neither minds nor bodies.

and argumentation" (EHU 11.2). "Pertinacious bigotry" is "fatal to philosophy" (EHU 11.3). "Implicit faith and security . . . is the bane of all reasoning and free enquiry" (EHU 4.3). Philosophers ought to be modest and quick to point out the difficulties in their own positions (EHU 4.14).

PHILOSOPHY AND PRACTICAL JUSTIFICATION

In the *Enquiry*, as in the *Treatise*, the norms and aims of philosophy are not practical or moral. Hume makes this explicit in Section 8:

> There is no method of reasoning more common, and yet none more blameable, than, in philosophical disputes, to endeavour the refutation of any hypothesis, by a pretence of its dangerous consequences to religion and morality. When any opinion leads to absurdities, it is certainly false; but it is not certain that an opinion is false, because it is of dangerous consequence. Such topics, therefore, ought entirely to be forborne; as serving nothing to the discovery of truth, but only to make the person of an antagonist odious. (EHU 8.26)

Philosophy as such aims at no practical goods other than accuracy and truth. His preferred "ACADEMIC or SCEPTICAL philosophy" mortifies every passion "except the love of truth; and that passion never is, nor can be carried to too high a degree" (EHU 5.1).

Because philosophy as such does not aim at practical goods (besides truth), it is an open question whether the practical liabilities of philosophy outweigh its practical benefits. Hume addresses this open question in Section 1. Philosophy does have practical benefits. The second species of moral philosophy (philosophy proper), like the first, "has its peculiar merit, and may contribute to the entertainment, instruction, and reformation of mankind" (EHU 1.1). It benefits man insofar as he is a "reasonable . . . being" by forming his understanding (EHU 1.2). As he puts it later, "Man is a reasonable being; and as such, receives from science his proper food and nourishment" (EHU 1.6).

Hume then compares the practical benefits and drawbacks of the first and second species of philosophy (EHU 1.3–6). In the course of this comparison, he seems to concede many drawbacks of the second species, and to grant the superior practical value of the first species. For example, he appears to concede that the first species of philosophy is "more agreeable" and "more useful" than the second (EHU 1.3). In fact, he is only reporting the opinion of "many," and the preferences of "the generality of mankind." However, he seems to be speaking in his own voice when he acknowledges that the first species "enters more into common life," whereas the second cannot do so. The first species of philosophy has acquired "the most durable, as well as the justest fame" (EHU 1.4). On the other hand, "The mere philosopher"—that is, the second species—"is a character, which is commonly but little acceptable in the world" (EHU 1.5). Hume does not endorse the popular supposition that the "mere philosopher" contributes "nothing either to the advantage or pleasure of society" (EHU 1.5). Neither does he endorse the supposition that the "most perfect character" lies between the two extremes of "the mere philosopher" and "the mere ignorant" (EHU 1.5). But Hume does seem to be speaking in his own voice when he says that "so narrow are the bounds of human understanding, that little satisfaction can be hoped for in this particular [i.e., science], either from the extent or security of his [man's] acquisitions" (EHU 1.6).

Hume concludes his survey of the practical criticisms of pure philosophy with a statement of what "seems" to be the verdict of "nature":

> It seems, then, that nature has pointed out a mixed kind of life as most suitable to human race, and secretly admonished them to allow none of these biasses to *draw* too much, so as to incapacitate them for other occupations and entertainments. Indulge your passion for science, says she, but let your science be human, and such as may have a direct reference to action and society. Abstruse thought and profound researches I prohibit, and will severely punish, by the pensive melancholy which they introduce, by the endless uncertainty in which they involve you, and by the cold reception which your pretended discoveries shall meet with, when communicated. Be a philosopher; but, amidst all your philosophy, be still a man. (EHU 1.6)

The pure pursuit of philosophy results in melancholy, uncertainty, and public indifference or hostility. These are admittedly drawbacks that can only be offset by mixing philosophy with social life and active pursuits.

Hume is glad to admit that philosophy has its drawbacks, but he insists that it has significant practical benefits as well that justify the enterprise. Against "the absolute rejecting of all profound reasonings, or what is commonly called *metaphysics*, we shall now proceed to consider what can reasonably be pleaded in their behalf" (EHU 1.7). In the first place, just as the science of anatomy helps the painter depict the human body, so the "accurate and abstract philosophy" of man helps the "easy and humane" moralists effectively portray human nature (EHU 1.8). Second, philosophy serves society by diffusing a "spirit of accuracy" over all arts and professions (EHU 1.9). Third, philosophy provides innocent and harmless pleasure to those disposed toward it (EHU 1.10). Fourth, we "must cultivate true metaphysics with some care, in order to destroy the false and adulterate" (EHU 1.11–12). Fifth, merely mapping the various "parts and powers of the mind" is a valuable enterprise (EHU 1.13–14). If a description of the positions and movements of the planets merited the labors of scientists, much more so does the delineation of the mind, "in which we are so intimately concerned" (EHU 1.14). Sixth, we may reasonably hope to discover something of the causal principles that explain the operations of the human mind (EHU 1.15). Hume admits that "the abstractedness of these speculations is no recommendation, but rather a disadvantage to them" (EHU 1.17). Nonetheless, such abstraction is necessary for discovering truths "in subjects of such unspeakable importance" (EHU 1.16). Furthermore, he seeks to minimize this difficulty by writing with as much "clearness" and "novelty" as possible (EHU 1.17).

Hume's conception of philosophy in the *Enquiry* holds few surprises for readers of the *Treatise*. It is still his key terminology for the normative method of the sciences. He can take for granted that the demands of philosophy as he conceives it are, in broad outline, familiar to anyone acquainted with experimental science. Philosophy endorses the same set of belief-forming processes as in the *Treatise*. Philosophy is independent of practical considerations, and has practical liabilities and benefits, when pursued moderately. As in the *Treatise*, the demands of philosophy do generate skeptical dilemmas, to which I turn in the next chapter.

Chapter Seven

Hume's Purely Practical Response to Skepticism in the *Enquiry*

In this chapter I review Hume's catalog of skeptical arguments and his response to them in Section 12 ("Of the Academical or Sceptical Philosophy") of the *Enquiry*. Hume puts forward basically the same arguments against sensory beliefs as he did in the *Treatise*. I argue against non-skeptical interpreters that he still regards these as sound arguments against a core belief. He puts forward different arguments against reason than he did in the *Treatise*. He does not, I argue, regard these as sound, but neither does he refute them. He uses them as *reductio* arguments to drive his readers to epistemic skepticism. Then he commends Academic philosophy, which provides a purely psychological and practical response to skeptical arguments, but not an epistemic one.

I examine each argument in the order of its appearance in Section 12. Although Hume previews his response to skepticism in Section 5.1–2, Section 12 functions as the correlative of *Treatise* 1.4.7. I take up Hume's famous discussion of causal inference from Sections 4–5 when I exposit the "*philosophical* objections" to probable reasoning (EHU 12.22).

ANTECEDENT SKEPTICISM: EXCESSIVE AND MODERATE

In the opening paragraphs of Section 12, Hume introduces the topic of the relationship between philosophical argument and psychological compulsion. Many theists present arguments against atheism while at the same time claiming that there are no "speculative atheists," no people who sincerely disbelieve in any deity (EHU 12.1). Willy-nilly, people believe in a god, no matter what they say, and regardless of arguments. But this makes theistic proofs look like works of epistemic supererogation. Hume seems to be suggesting that the divines themselves are in bad faith: if they really thought that there were no speculative atheists, then they would not continually attempt to argue for theism. We do not need to make arguments against positions that no one actually holds—even if such arguments are available.

Hume points out a parallel between speculative atheism and skepticism. First, "it is certain, that no man ever met with any such absurd creature, or conversed with a man, who had no opinion or principle concerning any subject, either of action or speculation" (EHU 12.2). The implication we might immediately draw from the preceding paragraph is that therefore we do not need to refute skepticism—even if refutations are available. I think this is exactly the conclusion Hume draws in this chapter. However, he does not immediately announce it. Instead, he proposes to take up two questions: "What is meant by a sceptic? And how far it is possible to push these philosophical principles of doubt and uncertainty?" (EHU 12.2).

Hume answers the first question, "What is a sceptic?", by introducing two cross-cutting divisions (antecedent/consequent, and moderate/excessive) that yield four varieties of skepticism.[1] He begins by discussing two varieties of antecedent skepticism, excessive (or Cartesian) and moderate (EHU 12.3–4). Then comes his exposition of arguments for consequent skepticism (EHU 12.5–22). As he ends this catalog, he transitions into a discussion of two forms of consequent skepticism, excessive (or Pyrrhonian) (EHU 12.21–25) and moderate (or Academic) (EHU 12.24–34). The second question, "how far it is possible to push these philosophical principles of doubt and uncertainty?", is psychological. As it turns out, both forms of excessive skepticism (Cartesian and Pyrrhonian) are psychologically impossible to achieve or sustain for long. They also cut off the essential beliefs of common life and philosophy. Both forms of moderate skepticism, however, support the practice of philosophy.

Hume also criticizes Cartesian skepticism on philosophical grounds. There is no "original principle," like the cogito ("I think, therefore I am"), "which has a prerogative above others, that are self-evident and convincing" (EHU 12.3). Neither is there any possible argument for the reliability of our faculties that does not already presuppose that very reliability. "The CARTESIAN doubt, therefore, were it ever possible to be attained by any human creature (as it plainly is not) would be entirely incurable; and no reasoning could ever bring us to a state of assurance and conviction upon any subject" (EHU 12.3). With these observations, Hume dismisses excessive antecedent skepticism. As noted in the last chapter, he embraces moderate antecedent skepticism as normative for philosophy as such (EHU 12.4).

---

1. Qu, *Hume's Epistemological Evolution*, 182.

ARGUMENTS FOR CONSEQUENT SKEPTICISM

Having treated antecedent skepticism, Hume now introduces consequent skepticism:

> There is another species of scepticism, *consequent* to science and enquiry, when men are supposed to have discovered, either the absolute fallaciousness of their mental faculties, or their unfitness to reach any fixed determination in all those curious subjects of speculation, about which they are commonly employed. (EHU 12.5)

Cartesian skepticism cuts off philosophical inquiry before it begins. But consequent skepticism begins with philosophical inquiry into human nature, taking for granted the legitimacy of philosophy's norms. The discoveries of this inquiry then provide the basis for a negative verdict on the faculties of human nature.

Consequent skepticism targets the core beliefs ("the maxims of common life") and faculties (he mentions "our very senses") that make common life and philosophy possible (EHU 12.5). Hume proposes to "enquire into the arguments, on which they [these paradoxical tenets] may be founded." The arguments in the skeptical catalog target the senses (EHU 12.6–16, the remainder of Part 1) and reason (Part 2, EHU 12.17–22), both demonstrative (EHU 12.18–20) and probable (EHU 12.21–22).

Trite Topics

The skeptical catalog begins with a class of arguments that Hume mentions only to dismiss (EHU 12.6). These are arguments from the "imperfection and fallaciousness" of our sensory organs on some occasions, to the conclusion that we should not trust our senses at all. Our senses do deceive us in cases of perceptual variation: "the crooked appearance of an oar in water; the various aspects of objects, according to their different distances; the double images which arise from the pressing one eye; with many other appearances of a like nature" (EHU 12.6).[2] But Hume rejects the inference to skepticism, and he does so on purely epistemic grounds:

2. Hume's crucial argument against the vulgar belief in the senses in the *Treatise* is precisely an argument from "the double images which arise from the pressing one eye" (cf. T 1.4.2.45). *Enquiry* 12.6 may then seem like a renunciation of the *Treatise* argument. However, he also argues against the vulgar belief a few paragraphs later from "the various aspects of objects, according to their different distances": "The table, which we see, seems to diminish, as we remove farther from it" (EHU 12.9). In short, though he dismisses perceptual variation arguments in EHU 12.6, he uses them in EHU 12.9 as well as in T1.4.2.45. It seems that, according to Hume, there are good arguments as well as bad arguments which can be made from precisely the same phenomena of perceptual variation. It is not necessary for me here to determine exactly how Hume distinguishes the good perceptual variation arguments from the bad ones.

> These sceptical topics, indeed, are only sufficient to prove, that the senses alone are not implicitly to be depended on; but that we must correct their evidence by reason, and by considerations, derived from the nature of the medium, the distance of the object, and the disposition of the organ, in order to render them, within their sphere, the proper *criteria* of truth and falsehood. (EHU 12.6)

The only reason Hume even broaches these "trite topics" is to assure us that the consequent skeptic need not lean on such a broken reed. "There are other more profound arguments against the senses, which admit not of so easy a solution" (EHU 12.6)

The Vulgar Belief in Body

Next, Hume reprises his arguments from the *Treatise* against three forms of belief in body—the vulgar belief, the double existence theory, and the modern philosophy. He begins by noting that humans universally and instinctively believe that the very perceptions present to their minds are "external objects" whose existence does not depend on their minds (EHU 12.7–8). Hume gives no indication that anything is wrong with this instinctive belief as such. It seems entirely plausible to assume that this universal belief is a belief of common life; it is the primitive form of experience on which all causal reasoning is based. After Hume puts forward a defeater argument for the instinctive belief, he writes that "philosophy . . . can no longer plead the infallible and irresistible instinct of nature" in support of belief in body (EHU 12.10). This sentence implies that prior to the defeater argument, "philosophy" *could* plead the "infallible and irresistible instinct of nature" in support of belief in body. (The instinct is not "infallible" in the sense of unerring, but in the sense that it always produces belief). The implication seems to be that "philosophy" approves of the infallible and irresistible instinct of nature, and recognizes the products of such instinct as defeasibly justified beliefs.

However, the instinctive belief in body faces a defeater argument from causal reasoning.

> The table, which we see, seems to diminish, as we remove farther from it: But the real table, which exists independent of us, suffers no alteration: It was, therefore, nothing but its image, which was present to the mind. (EHU 12.9)

I take it that Hume's compressed argument runs as follows:

1. Our perceptions change when we move. ("The table, which we see, seems to diminish, as we remove farther from it.")

2. If perceptions are identical with objects, then objects also change when we move. (This suppressed premise follows from the Indiscernibility of Identicals and 1.)
3. But objects do not change when we move. ("But the real table, which exists independent of us, suffers no alteration.")[3]
4. Therefore, our perceptions are not identical with objects. ("It was, therefore, nothing but its image, which as present to the mind.")

Thus, "the slightest philosophy," "the obvious dictates of reason," overturn the instinctive vulgar belief in body (EHU 12.9).

The System of Double Existences

In the first *Enquiry*, as in *Treatise* 1.4.2, Hume claims that reflection on the defeat of the vulgar belief naturally produces a belief in the system of double existences: "So far, then, are we necessitated by reasoning to contradict or depart from the primary instincts of nature, and to embrace a new system with regard to the evidence of our senses" (EHU 12.10). According to this new system, "the perceptions of the mind must be caused by external objects, entirely different from them, though resembling them" (EHU 12.11).

The system of double existences has no source of philosophical justification, since it derives neither from "the infallible and irresistible instinct of nature," nor from any possible rational argument.

> But here philosophy finds herself extremely embarrassed, when she would justify this new system, and obviate the cavils and objections of the sceptics. She can no longer plead the infallible and irresistible instinct of nature: For that led us to a quite different system, which is acknowledged fallible and even erroneous. And to justify this pretended philosophical system, by a chain of clear and convincing argument, or even any appearance of argument, exceeds the power of all human capacity. (EHU 12.10)

The next four paragraphs argue that the system of double existences has nothing to recommend it to reason (EHU 12.11–14). This part of the argument runs slightly differently than in the *Treatise*. First, in the *Treatise*, the argument against the vulgar belief did not presuppose the existence of any continued and distinct existences. So Hume could criticize the adherents of the double existence theory because they "arbitrarily invent a new set of perceptions," never present to any mind, which they construe as objects (T 1.4.2.56). But in the *Enquiry*, the argument against the vulgar belief assumes

3. I think Hume intends to make basically the same argument here as in T 1.4.2.45. However, it is worth noting that premise 3 makes no appearance in T 1.4.2.45.

as a premise that there are mind-independent objects (see premise 3 above). So Hume cannot (and does not) criticize the double existence theory on the grounds that this assumption is arbitrary.

Instead of criticizing the assumption that there are bodies that cause our perceptions, Hume focuses his attack on the gratuitous assumption that these objects have qualities resembling our perceptions. As in the *Treatise*, Hume's fundamental criticism of any possible argument for a qualitative resemblance between bodies and our perceptions is that we have no experiential basis for such an argument (EHU 12.12; cf. T 1.4.2.47). But Hume adds new material that engages with Descartes's *Meditations* in several ways too. Two lines of argument appropriate the skeptical Descartes of the First Meditation. First, Hume raises the possibility that our own minds or some "unknown spirit" might cause our perceptions (EHU 12.11). Second, we commonly assume that in the cases of "dreams, madness, and other diseases," our perceptions are not caused by resembling external objects (EHU 12.11). If this assumption is true, it provides evidence against the double existence theory.

Other lines of argument implicitly attack the anti-skeptical Descartes of the later Meditations. First, if mind-body dualism (like Descartes's) is true, then it is "inexplicable" how a body could "convey an image of itself" to a mind (EHU 12.11). Second, contra Descartes, we cannot call in a Supreme Being to guarantee the reliability of our senses (EHU 12.13). Hume gives two reasons why this move fails. The first is an argument Descartes himself puts forward in the First Meditation: "If his [God's] veracity were at all concerned in this matter, our senses would be entirely infallible; because it is not possible that he can ever deceive" (EHU 12.13). But our senses are not infallible, so God's veracity has no bearing on the reliability of our senses. Hume has no time for the answer Descartes gives to this argument in later Meditations. Second, if we do not presuppose the veracity of our senses, then we cannot argue from the existence of the external world to the existence of a non-deceiving God. Hume assumes that such an argument is the only source of justified belief in God. We need justified sensory beliefs to ground belief in God, so we cannot invoke God to justify our sensory beliefs. (Hume tacitly sidesteps Descartes's arguments in the Third and Fifth Meditations from the mere idea of God to his existence). All told, the system of double existence has no recommendation either to nature or reason. "This is a topic, therefore, in which the profounder and more philosophical sceptics will always triumph" (EHU 12.14).[4]

4. Hume omits from the *Enquiry* the defeater argument that he presents in *Treatise* 1.4.4.3–5, the argument that leads from the double existence theory to its refinement, the modern philosophy. He also omits any mention of the "antient philosophy" of substance metaphysics (cf. T 1.4.3).

Modern Philosophy

Next, Hume presents a defeater argument for the modern philosophy—that is, the belief that the sensible "secondary" qualities exist only in the mind, and resemble nothing in the object, which has only the primary qualities of extension and solidity (EHU 12.15). As in the *Treatise*, he gives a Berkeleian argument. In the *Treatise*, he devotes nine paragraphs to this argument (T 1.4. 4.6–14); in the *Enquiry*, he compresses it into one. In sum:

> The idea of extension is entirely acquired from the senses of sight and feeling; and if all the qualities, perceived by the senses, be in the mind, not in the object, the same conclusion must reach the idea of extension, which is wholly dependent on the sensible ideas or the ideas of secondary qualities. (EHU 12.15)

The only way to skirt this argument is to invoke the notion of "*Abstraction*" that Berkeley has exploded. Thus, if secondary qualities are in the mind only, then so too are primary qualities. The modern philosophy bereaves matter "of all its intelligible qualities" and leaves us "only a certain unknown, inexplicable *something*, as the cause of our perceptions; a notion so imperfect, that no sceptic will think it worth while to contend against it" (EHU 12.16). Hume concludes that all these accounts of "the evidence of sense" or "the opinion of external existence" face insuperable philosophical objections (EHU 12.16).

Hume gives no epistemic answer to any of the skeptical arguments against belief in senses. It does not seem as though any epistemic response is available to him, given the norms of philosophy. He can only save the vulgar belief by rejecting the authority of causal reasoning. He can only save the double existence theory by countenancing factual beliefs that are unsupported either by natural instinct or by experience-based causal reasoning. Saving the modern philosophy would require a modification of Hume's fundamental theory of ideas.

In the *Treatise*, the Title Principle permits Hume to ignore the authority of causal reasoning when psychological and practical necessity requires it. This allows him to ignore the defeater argument against the vulgar belief. Hume hints at the relevance of practical considerations when he mentions that "arguments and reasonings" against the senses "can so little serve to any serious purpose" (EHU 12.15). He hints at the relevance of psychological considerations (and the absence of epistemic solutions) when he says that Berkeley's arguments "*admit of no answer and produce no conviction*" (EHU 12.15, footnote 32).

Qu does not deny the force of these skeptical arguments against body, but he denies that they target what Hume regards as a core belief. Qu sees the "more profound" arguments against the senses as successfully showing that we have no justification for believing in an external, mind-independent universe.[5] The mitigated skeptic therefore refrains from holding such beliefs. This concession might seem to cut off all of the beliefs of common life and science, but Qu thinks that Hume still leaves room for metaphysically neutral beliefs about mere appearances. "In particular, beliefs regarding the continued and distinct existence of objects would be in order when understood in terms of the realm of appearances, and illegitimate otherwise."[6] Thus, belief in mind-independent objects is unjustified, but also unnecessary.

I do not think Hume can hold this position. First, he never suggests that belief in continued and distinct existences is dispensable, psychologically or practically. In the *Treatise* he stresses that no one can suspend belief in continued and distinct existences for long, regardless of skeptical objections. Hume makes a similar point in the *Enquiry*. He says of Berkeley's skeptical arguments on this head that though "*they admit of no answer*," yet they "*produce no conviction*. Their only effect is to cause that momentary amazement and irresolution and confusion, which is the result of scepticism" (EHU 12.15, footnote 32.1). So on Qu's reading, even the Academic will be psychologically compelled to hold unjustified beliefs in bodies.

Second, I do not think that continued and distinct existences can be "understood in terms of the realm of appearances" by Hume's lights. Qu does not spell out exactly what this could mean for Hume, and he gives no textual evidence for it. Millican suggests that "apparently there is no harm in thinking of them [external objects] using the merely relative idea of 'a certain unknown, inexplicable *something*, as the cause of our perceptions,'" as Hume puts it (EHU 12.16).[7] But this is exactly the notion Hume derides as "so imperfect, that no sceptic will think it worth while to contend against it" (EHU 12.16). Millican suggests that this statement is intentionally ambiguous, not damning, as it sounds. Hume may feel that the threadbare notion of objects serves his "irreligious purposes and his desire to cripple any ambition towards a rival metaphysics based on supposed rational insight into the nature of matter."[8] But Hume does not give any explicit indication of these motives, and his derision of the "inexplicable *something*" seems sincere.

Garrett follows a different non-skeptical interpretive strategy. On Garrett's reading, the skeptical arguments in Section 12 of the *Enquiry* (and in *Treatise* 1.4.7) are "doubt-inducing discoveries."[9] Hume's arguments about bodies in particular lower the probability that sensory beliefs are true.[10]

5. Qu, *Hume's Epistemological Evolution*, 192–93.
6. Ibid. 192n22. He seems to follow Millican's similar line of thought. Millican, "Hume's Chief Argument," 99, 100.
7. Millican, "Hume's Chief Argument," 99
8. Ibid., 100
9. Garrett, *Hume*, 213.
10. Ibid., 220–22.

However, they do not lead to the conclusion that sensory beliefs are unjustified. Instead, a process of reflection on the doubt-inducing considerations leads Hume retain his trust in reason and the senses, albeit with lower confidence levels befitting a mitigated skeptic.[11]

Garrett's sophisticated reading of Hume is difficult to square with the text though. If Hume thinks belief in body is justified, then he either thinks that the vulgar belief, the double existence theory, or the modern philosophy is justified. The vulgar belief is supported by natural instinct, but defeated by reason. The double existence theory is supported neither by natural instinct nor by reason. The modern philosophy is unintelligible. So Hume believes either that an irrational belief is justified (the vulgar), a belief unsupported by instinct or reason is justified (the double existence theory), or that an unintelligible belief is justified (the modern philosophy). None of these options seem plausible. If Hume ultimately revises his assessment of reason's defeat of the vulgar belief, or reason's non-support of the double existence theory, he never mentions it.

Abstract Reasoning

Hume uses the absurdity of the infinite divisibility of quantity as an argument against the reliability of abstract reasoning in general. Abstract reasoning deals with quantity and number (cf. EHU 4.1, 12.27). The quantities Hume discusses here are quantities of extension or space (EHU 12.18) and quantities of time (EHU 12.19). Geometry is the abstract science dealing with quantities of space or extension. Hume lays it down as indisputable that the infinite divisibility of extension is unintelligible and absurd. But the demonstrative arguments for infinite divisibility seem completely irreproachable:

> These seemingly absurd opinions are supported by a chain of reasoning, the clearest and most natural; nor is it possible for us to allow the premises without admitting the consequences . . . The demonstration of these principles seems as unexceptionable as that which proves the three angles of a triangle to be equal to two right ones, though the latter opinion be natural and easy, and the former big with contradiction and absurdity. (EHU 12.18)

The irreproachability of the demonstrative argument for an absurd conclusion yields skeptical paralysis:

> Reason here seems to be thrown into a kind of amazement and suspence, which . . . gives her a diffidence of herself . . . she can scarcely pronounce with certainty and assurance concerning any one object. (EHU 12.18)

11. Ibid., 227–37.

I summarize the argument as follows:

1. If there is a good demonstrative argument for the unintelligible thesis that quantity is infinitely indivisible, then abstract reasoning does not justify its conclusions.
2. But there is a good demonstrative argument for the unintelligible thesis that quantity is infinitely indivisible.
3. Therefore, abstract reasoning does not justify its conclusions.

Hume not only has a satisfying philosophical, epistemic answer to this argument available, but he actually tells us what the answer is. Hume rejects premise 2 of the argument. In a footnote, he says "It seems to me not impossible to avoid these absurdities and contradictions"; Hume has on hand "the readiest solution of these difficulties," which he briefly references (EHU 12.20n34).[12] In *Treatise* Book 1, Part 2 he develops his theory of the finite divisibility of space and time at length, including rebuttals to the arguments of the mathematicians (T 1.2.2.7–10; 1.2.4–5). As I noted in chapter 1, Hume holds that geometrical demonstrations lose their force when they deal with very minute objects, so they cannot prove the infinite divisibility of extension.

Once again, Hume is making a *reductio* skeptical argument that does not arise for someone with his philosophical commitments. This explains why Hume would advance a skeptical argument to which he has the solution. When he mentions the "trite topics," he does so in order to dismiss these objections. But he does not dismiss this argument against abstract reason. He put his answer to the "trite topics" in the main body of the text. But he puts his epistemic solution to the argument against abstract reason in a footnote, indicating that it is not essentially relevant to his main point. That main point is that we have practical reasons for retaining most beliefs produced by valid abstract reasoning, no matter what skeptical arguments can be brought against them.

---

12. Fogelin rightly notes that "if Hume is right in saying that the doctrine of infinite divisibility is avoidable, then . . . no general skepticism with regard to reason is forthcoming." Fogelin, *Hume's Skeptical Crisis*, 152. Garrett notices the solution in the footnote, but still tries to construe the argument for infinite divisibility as inducing doubt about reason: "Even if ultimately soluble, he [Hume] suggests, the naturalness and difficulty of such paradoxes offers a basis for some lowering of the probability that attempted abstract reasoning—that is, attempted demonstration—is veracious." Garrett, *Hume*, 227. However, the fact that many fallacious arguments are initially attractive and beguiling does not seem like a serious cause for worry. Surprisingly, some commentators on Part 2 of EHU 12 simply make no mention of the footnote. See Stephen Buckle, *Hume's Enlightenment Tract: The Unity and Purpose of* An Enquiry concerning Human Understanding (Oxford: Clarendon Press, 2001), 308–9; David Fate Norton, "Of the Academical or Sceptical Philosophy," in *Reading Hume on Human Understanding*, ed. Peter Millican (Oxford: Clarendon Press, 2002), 376–77.

Popular Objections to Probable Reasoning

Instead of presenting the skeptical argument against reason from *Treatise* 1.4.1, he patches together three other bits from Book 1 of the *Treatise* and now repurposes them as skeptical arguments. The argument from *Treatise* 1.4.1 manages to undermine belief in demonstrative conclusions as well as belief in probable conclusions. In the absence of this argument Hume must present separate arguments to target "abstract" reasoning and inductive inference respectively. Why would Hume replace the argument from *Treatise* 1.4.1 with these other two *reductio* skeptical arguments? A rhetorical explanation is readily available. The argument from *Treatise* 1.4.1 is very abstruse and refined (T 1.4.1.10–11; 1.4.7.7). Hume wants to make more widely accessible arguments.[13]

The first kind of skeptical objection to probable reasoning is "popular":

> The popular objections are derived from the natural weakness of human understanding; the contradictory opinions, which have been entertained in different ages and nations; the variations of our judgment in sickness and health, youth and old age, prosperity and adversity; the perpetual contradiction of each particular man's opinions and sentiments; with many other topics of that kind. (EHU 12.21)

It is hard to know exactly how this this highly compressed argument should be construed. Possibly, Hume has in mind a simpler argument such as the following:

1. If probable reason often produces contradictory results, then it does not defeasibly justify beliefs.
2. But probable reason does often produce contradictory results.
3. Therefore, probable reason does not defeasibly justify beliefs.

It seems as though Hume has a perfectly satisfying philosophical answer to this argument available. The philosopher should reject premise 1. There are multiple reasons why probable reason might produce contradictory beliefs. If we take "probable reason" in a broad sense, then it is a faculty that produces

13. Garrett leaves it open whether Hume drops the argument simply because he "judged these topics unnecessary and too complex for the later and more streamlined work" or because he lost confidence in its soundness. Garrett, *Hume*, 227. Against the view that Hume drops the argument because he has lost faith in its soundness, I note that he puts something very similar to it in the mouth of Philo in Part 1 of the *Dialogues*: "All sceptics pretend, that, if reason be considered in an abstract view, it furnishes invincible arguments against itself, and that we could never retain any conviction or assurance, on any subject, were not the sceptical reasonings so refined and subtile, that they are not able to counterpoise the more solid and more natural arguments, derived from the senses and experience" (D 1.11). Cf. Owen, "Scepticism with Regard to Reason," 109.

both good and bad inductive inferences. Bad inductive inferences are unsupported by adequate experiential evidence, and sadly, humans make them frequently. Good and bad inductive inferences may conflict with each other. But just because philosophers reject the defeasibly justifying power of bad inductive inferences does not require them to reject the defeasibly justifying power of good ones. Sometimes, one good inductive inference is defeated by a better one that is based on more experience—as in the case of the Indian prince who discovered that in fact water freezes at low temperatures (cf. Section 10). The philosophers do not reject probable reason on this basis though. Probable reason can still produce defeasibly justified beliefs, even if some of those beliefs later meet with epistemic defeat from observation or from more probable reasoning.

We could also construe the "popular objections" in terms of reliability, as follows:[14]

1. If probable reason produces contradictory beliefs, then it is unreliable.
2. But probable reason does produce contradictory beliefs.
3. Therefore, probable reason is unreliable.
4. If probable reason is unreliable, then it is not defeasibly justifying.
5. Therefore, probable reason is not defeasibly justifying.

This construal of the popular objections seems to invite refutation as well. First, premise 1 may be false. Possibly, probable reason produces some contradictory beliefs, but still produces mostly true beliefs, making it reliable if not infallible. Premise 1 also invites another response: perhaps reason is less reliable in some circumstances than others. It deserves confidence in each kind of circumstance proportionate to its observed reliability in those circumstances.

Despite the fact that Hume has epistemic answers to the popular objections available, he does not state them. Instead, he writes the following:

> It is needless to insist farther on this head. These objections are but weak. For as, in common life, we reason every moment concerning fact and existence, and cannot possibly subsist, without continually employing this species of argument, any popular objections, derived from thence, must be insufficient to destroy that evidence. The great subverter of *Pyrrhonism* or the excessive principles of scepticism, is action, and employment, and the occupations of common life. These principles may flourish and triumph in the schools; where it is, indeed, difficult, if not impossible, to refute them. But as soon as they leave the shade, and by the presence of the real objects, which actuate our passions and

14. I do not think that Hume would endorse Premise 4, which ties defeasible justification to reliability.

> sentiments, are put in opposition to the more powerful principles of our nature, they vanish like smoke, and leave the most determined sceptic in the same condition as other mortals. (EHU 12.21)

His only reply to the objections is that, in common life, we are psychologically incapable of suspending our belief in the deliverances of reason.[15] The popular objections cannot destroy the evidentness, the psychological strength, of our probable reasonings, in common life.

The fact that the skeptic's arguments are psychologically weak does not mean that they are epistemologically defective. On the contrary, as Hume notes, "These principles [of excessive skepticism] may flourish and triumph in the schools; where it is, indeed, difficult, if not impossible, to refute them." Skeptical principles "flourish and triumph" psychologically when we give up our core beliefs. The "flourishing" and "triumphing" of the principles of skepticism are here directly connected with the difficulty, if not impossibility, of philosophically refuting them. Hume does not flatly state that skepticism is irrefutable in the schools, but he does suggest it as a possibility. So his response to Pyrrhonian skepticism does not depend on having a refutation to skeptical arguments. Furthermore, it is not as though refutations to skepticism are easier to think of in common life than they are in the schools. Pyrrhonism fares better in schools than outside of them just because it is harder to think about *any* arguments, skeptical or anti-skeptical, while engaged in active life. Hume's point is that in common life we find it impossible to resist the powerful principles of our nature any longer. Discredited or not, we instinctively resume our core beliefs.[16] Pyrrhonism can psychologically triumph where, but only where, we are in a position to focus exclusively on skeptical arguments.

Philosophical Objections to Probable Reasoning

Hume's second skeptical argument against probable reasoning purports to be more serious than the popular objections.

> The sceptic, therefore, had better keep within his proper sphere, and display those *philosophical* objections, which arise from more profound researches. Here he seems to have ample matter of triumph. (EHU 12.22)

15. Qu agrees that at this point Hume raises only "pragmatic concerns against Pyrrhonian scepticism as a whole, noting that it is detrimental as well as psychologically untenable (EHU 12.23)." Qu, *Hume's Epistemic Evolution*, 181.

16. This is the same point that Hume makes in *Treatise* 1.4.1, suggesting again that he may be reprising that material here in compressed form. It would be surprising though if he now reclassified that highly abstruse argument as a "popular objection." Furthermore, the popular objections are aimed only at probable reasoning, whereas *Treatise* 1.4.1 targets demonstrative and probable reasoning.

The difference between "popular" and "philosophical" skeptical arguments has to do with the contexts in which they are likely to be appreciated. Even the vulgar could understand the "popular" objection that people often reason to contradictory conclusions. But it takes considerable philosophical reflection to grasp the point that there is no noncircular argument for the Uniformity Principle, and that causal inference is produced by custom. (This is the specific "philosophical" objection that Hume is about to take up). This argument is therefore better suited to the "proper sphere" of the skeptic, the "schools," the sphere where subtle arguments can be made and contemplated. It is precisely here that the skeptic can triumph—that is, where the skeptic may successfully cause us to suspend our core beliefs for a time. "While the sceptic insists upon these topics, he shews his force, or rather, indeed, his own and our weakness; and seems, for the time at least, to destroy all assurance and conviction" (EHU 12.22).

The skeptical argument comprises one long sentence:

> He [the skeptic] justly insists, that all our evidence for any matter of fact, which lies beyond the testimony of sense or memory, is derived entirely from the relation of cause and effect; that we have no other idea of this relation than that of two objects, which have been frequently *conjoined* together; that we have no argument to convince us, that objects, which have, in our experience, been frequently conjoined, will likewise, in other instances, be conjoined in the same manner; and that nothing leads us to this inference but custom or a certain instinct of our nature; which it is indeed difficult to resist, but which, like other instincts, may be fallacious and deceitful. (EHU 12.22)

Hume concedes to the skeptic that the causal inference is produced only by fallible custom, not argument. This is the claim that Hume himself defends in Sections 4–5. Now he suggests that this claim somehow raises a skeptical problem. But aside from this one premise, the skeptical "argument" is left implicit, at most suggested. The implicit argument seems to run as follows:

1. If causal inference is produced by fallible custom rather than reason, then it is not defeasibly justifying. (Unstated)
2. But causal inference is produced by fallible custom rather than reason. (Stated)
3. Therefore, it is not defeasibly justifying. (Unstated)

Does Hume regard this argument as sound?

Considerable evidence seems to indicate that Hume regards this skeptical argument as sound.[17] To begin with, Hume makes no epistemic criticism of the philosophical objection to causal inference. Instead, he brushes the topic aside with the remark, "These arguments might be displayed at greater length, if any durable good or benefit to society could ever be expected to result from them" (EHU 12.22). Second, Hume's acceptance of the argument explains why he seems to forcefully endorse epistemic skepticism in Section 12, especially immediately following this passage. Third, the titles of Sections 4–5, where Hume first introduces the key claims about the source of causal inference, indicate that they have "Sceptical" implications. In short, these titles suggest or imply the same skeptical argument against causal inference, and reinforce the impression that Hume accepts that argument.

Fourth, Hume gives his analysis of causal inference a more epistemic turn in the *Enquiry* than in the merely descriptive *Treatise*.[18] In the *Treatise*, Hume emphasizes the merely descriptive claim that the supposition of the UP cannot *cause* itself: "The same principle cannot be both the cause and effect of another" (T 1.3.6.7). That is, he rejects a causal circle, the self-causation of a supposition. In the *Enquiry*, he emphasizes that the UP cannot *prove or support* itself; he rejects a circular argument, the self-justification of a supposition. He writes:

> all our experimental conclusions proceed upon the supposition, that the future will be conformable to the past. To endeavour, therefore, the proof of this last supposition by probable arguments, or arguments regarding existence, must be evidently going in a circle, and taking that for granted, which is the very point in question. (EHU 4.19)

Winkler thinks this provides evidence that Hume intends his argument to show that inductive expectations are not justified by reason.[19]

If Hume accepts the skeptical argument against causal inference in Section 4 and Section 12 as sound, that only supports my overall interpretation. On my reading, he is an epistemic skeptic about core beliefs, including the prod-

---

17. For example, according to Kenneth Winkler, Hume's point is that since the Uniformity Principle cannot be justified by reason, our inductive expectations are not justified. Winkler, "Hume's Skeptical Logic of Induction," in *The Oxford Handbook of Hume*, edited by Paul Russell (New York: Oxford University Press, 2016), 214–19.

18. Qu argues in detail that Hume's arguments in Sections 4 and 5 of the *Enquiry* are both more epistemologically normative than their counterparts in the more purely descriptive *Treatise*. On one hand, the argument in Section 4 of the *Enquiry* does appear to threaten at least one plausible source of justification for induction. On the other hand, Qu finds that Hume's treatment of custom in Section 5 does not merely explain the cause of inductive inferences, but shows the true source of their epistemic justification. So Hume's treatment of inductive inference in Sections 4–5 does not come to an ultimately skeptical conclusion. Qu, *Hume's Epistemological Evolution*, chapters 4–5.

19. Winkler, "Hume's Skeptical Logic," 217.

ucts of reason. His purely practical response to epistemic skepticism would apply to inductive skepticism as well as to any other kind.

However, I do not think that Hume actually does accept the argument that he suggests or implies in *Enquiry* 12.22 and Section 4. First, it is striking that Hume never—either in Section 4 or Section 12—explicitly endorses the unstated premise or conclusion of the argument that he suggests. He only explicitly endorses the descriptive premise of the argument, not the normative epistemic parts of it.[20] In both Section 4 and Section 12, the normative epistemic premise and conclusion are at most suggested indirectly, by implication. Second, in the *Treatise*, Hume denies the key epistemic premise of the argument. No skeptical problem arises from the same descriptive claims in *Treatise* 1.3.6. Philosophy endorses all PIU principles of the imagination, including custom, as defeasibly justifying (T 1.4.4.1). Of course, Hume might have a far more rigorist stance in the *Enquiry* than in the *Treatise*. But this would be a significant change in his philosophical principles, on a scale that seems to run contrary to his statements about the relation between the two works (see chapter 6).

Third, there is a plausible rhetorical explanation for why Hume would suggest this argument without refutation, even if he does not accept it himself. Once again, Hume is suggesting a *reductio* skeptical argument that arises for someone who accepts the premise that "If causal inference is produced by fallible custom rather than reason, then it is not defeasibly justifying." In fact, this is the same *reductio* argument that he has already made in *Treatise* 1.4.7.3 and Abstract 27. In *Treatise* 1.4.7.3, this *reductio* argument against probable reason supplements the more abstruse one from *Treatise* 1.4.1 that produces the "dangerous dilemma" in *Treatise* 1.4.7.7. In the Abstract and in the first *Enquiry*, the *reductio* argument replaces the more abstruse one. Dropping an abstruse argument in favor of a somewhat simpler one makes the *Enquiry* more readable than the *Treatise*, which is one of Hume's goals. Even if the argument against causal reasoning is not one which Hume himself accepts, it serves the rhetorical purpose of motivating readers to accept his purely practical response to philosophical (epistemic) skepticism.

Fourth, the *reductio* reading of the argument against probable reason is made more likely by the fact that Hume's other skeptical arguments against reason are also *reductio* arguments. The abstruse argument of *Treatise* 1.4.1 defeated both demonstrative and probable reasoning. In its absence, Hume needs another argument against demonstrative reasoning. He supplies it in the form of the paradox of infinite divisibility, as discussed above. This argument too is clearly intended as a *reductio* argument. Hume not only has a philo-

20. Winkler admits that Hume never actually says that since the Uniformity Principle cannot be justified by reason, our inductive expectations are not justified. Ibid., 217, 219.

sophical, epistemic solution available, but states it explicitly. By putting the epistemic solution in a footnote, he signals that it is not relevant to his main point, which is the practical solution to skepticism. The popular objections to causal reasoning also appear to be *reductio* skeptical arguments if, as I have suggested, Hume has good philosophical answers available (which he does not put forward). It is plausible to think that the philosophical objection to probable reason follows the same pattern as the other skeptical objections to reason.

Before moving on from the implicit *reductio* skeptical argument against causal inference, it is worth noting that Hume also uses it in Section 4 to motivate the acceptance of a purely psychological, practical response to skepticism in the first two paragraphs of Section 5. The section begins with a defense of "the Academic or Sceptical philosophy" on moral and prudential grounds from its dogmatic critics. This philosophy is morally edifying and "in almost every instance, must be harmless and innocent" (EHU 5.1). Hume anticipates the worry that Academic philosophy might "undermine the reasonings of common life, and carry its doubts so far as to destroy all action, as well as speculation" (EHU 5.2). His response is telling. He does not give any epistemic reason why Academic philosophy shields "the reasonings of common life" from all-encompassing, paralyzing skeptical objections. Instead, he simply appeals to the psychological force of our natural core beliefs in the face of any argument.

> Nature will always maintain her rights, and prevail in the end over any abstract reasoning whatsoever. Though we should conclude, for instance, as in the foregoing section, that, in all reasonings from experience, there is a step taken by the mind, which is not supported by any argument or process of the understanding; there is no danger, that these reasonings, on which almost all knowledge depends, will ever be affected by such a discovery. (EHU 5.2)

The Academic philosophy acknowledges the rights of nature; it approves and submits to the psychological power of natural core beliefs, regardless of whether these beliefs can be discredited by philosophical arguments. So far forth, Hume gives a purely practical response to the problem of induction. The purely practical interpretation provides an excellent explanation for why Hume should give Section 5 the seemingly paradoxical title, "Sceptical Solution of these Doubts." "Sceptical" refers to the fact that his solution is purely psychological and practical, and does not mitigate any of the philosophical force of the skeptical arguments. If he had a philosophical solution of these doubts, the solution would no longer be "sceptical."

After dismissing the philosophical objections to inductive reasoning on practical grounds, Hume launches back into a critique of excessive skepticism in general (EHU 12.23). The critique is twofold. Pyrhonnian skepticism is psychologically unsustainable and practically destructive. "But a Pyrrhonian cannot expect, that his philosophy will have any constant influence on the mind: Or if it had, that its influence would be beneficial to society" (EHU 12.23).[21] As for the preceding catalog of skeptical arguments, Hume does not even mention them, let alone refute their premises.

Hume closes out his practical rejection of Pyrrhonism by conceding its epistemic force. These objections

> can have no other tendency than to show the whimsical condition of mankind, who must act and reason and believe; though they are not able, by their most diligent enquiry, to satisfy themselves concerning the foundation of these operations, or to remove the objections, which may be raised against them. (EHU 12.23)

The Pyrrhonian objections may never produce conviction or practical good, but they can expose the fact that human life is based on ultimately unjustified beliefs.

Qu argues that we cannot take this statement of skeptical defeat at face value, since Hume obviously has refutations available for at least some of the skeptic's objections.[22] Qu acknowledges that "The strongest support for the sceptical reading of EHU 12 unquestionably derives from Hume's seeming concession [at EHU 12.14; 12.15, footnote 32; 12.21; 12.23] that the Pyrrhonian arguments have no answer."[23] But in *Enquiry* 12.6–23, Hume speaks from within the perspective of Pyrrhonism itself.[24] From within the assumptions and methodology of Pyrrhonism, the arguments are irrefutable, but not from Hume's own perspective.

But although Hume has refutations available for some of the skeptical arguments he presents in *Enquiry* 12, he does not have refutations available for all of them. At a minimum, the arguments for skepticism about external objects succeed by his own lights. His statements in *Enquiry* 12.14 and 12.15, footnote 32 reflect that fact. If even just one argument for extreme skepticism is irrefutable, then to that extent, extreme skepticism is irrefutable. This explains why Hume says so in *Enquiry* 12.23. Moreover, it seems hard to believe that everything Hume says about Pyrrhonism in *Enquiry* 12.6–12.23 is spoken from within the perspective of Pyrrhonism, rather than his own perspective. Within this stretch, he dismisses Pyrrhonism as psychologically

21. Similarly, Cleanthes criticizes Pyrrhonism as the psychologically impossible attempt to permanently reject discredited core beliefs (DNR 1.6). The attempt is not only futile, but practically unjustified: "And for what reason impose on himself such a violence? This is a point in which it will be impossible for him ever to satisfy himself, consistent with his sceptical principles" (DNR 1.6).

22. Qu, *Hume's Epistemological Evolution*, 206–7

23. Ibid., 205.

24. Ibid., 207

unsustainable, practically suicidal, and comically misguided. If we accept these statements as Hume's own opinion, we should also accept his statements about the epistemic strength of Pyrrhonism in the same context.

CONSEQUENT SKEPTICISM: HOW THE EXCESSIVE PRODUCES THE MODERATE

In Part 3, Hume continues to contrast two forms of consequent skepticism: excessive (Pyrrhonian) and mitigated (Academic). Already we know that Pyrrhonism responds to skeptical arguments against core beliefs by rejecting those beliefs. Hume has rejected Pyrrhonism on the grounds that it is psychologically unsustainable and practically suicidal. Already we know that Academic philosophy endorses natural core beliefs even in the face of irrefutable skeptical arguments (EHU 5.1–2). That is also clear, mainly by implication, from Section 12. Hume endorses Academic philosophy on practical grounds. Hume opens Part 3 by emphasizing the psychological benefits of Academic philosophy: it "may be both durable and useful" (EHU 12.24). Part 3 explains how an encounter with Pyrrhonism can actually produce two aspects of Academic skepticism (Hume calls these two aspects "species" of mitigated skepticism). These two aspects are intellectual modesty and restriction to the scope of common life.[25] Both of these aspects of Academic philosophy are already normative constituent aspects of philosophy *simpliciter*. The encounter with Pyrrhonism simply gives the Academic further motivation to follow the norms of philosophy *simpliciter*. All told, Academic philosophy endorses the norms of philosophy *simpliciter* except when those norms demand the suspension of core beliefs.

The first aspect of Academic skepticism is "a degree of doubt, and caution, and modesty, which, in all kinds of scrutiny and decision, ought for ever to accompany a just reasoner" (EHU 12.24). As early as Section 5, Hume says that "The academics always talk of doubt and suspense of judgment, of danger in hasty determinations" (EHU 5.1). This general intellectual modesty is not a distinctive norm of the Academics. It is materially the same as the moderate antecedent skepticism which is "a necessary preparative to the study of philosophy" in general (EHU 12.4). The consequent skeptical arguments of Pyrrhonism can, however, provide further motivation for adopting this attitude. Haughty dogmatists can learn from Pyrrhonism that the human mind at its best is afflicted with "strange infirmities," "universal perplexities and confusion" (EHU 12.24). These afford plenty of grounds for intellectual hu-

25. Hume does not say that a brush with Pyrrhonism is *necessary* for acquiring intellectual modesty or for confining our enquiries within the bounds of the understanding.

mility. Note that Pyrrhonism engenders intellectual humility just insofar as its we take its destructive arguments to succeed, philosophically. The Academic reasons that the Pyrrhonists are right, that even the beliefs of common life are not philosophically justified. Although we psychologically must and practically ought to continue to hold these beliefs, we ought to do so with humility and diffidence, in light of their ultimate philosophical status.

The second aspect is "the limitation of our enquiries to such subjects as are best adapted to the narrow capacity of human understanding" (EHU 12.25). Section 5 foreshadows this aspect too: "The academics always talk . . . of confining to very narrow bounds the enquiries of the understanding, and of renouncing all speculations which lie not within the limits of common life and practice" (EHU 5.1). Again, this scope restriction is not unique to Academic philosophy; it belongs to philosophy *simpliciter*, as Hume emphasizes from Section 1 onward. The brush with Pyrrhonism does not give the Academic a new scope restriction, but a new reason to abide within the same scope restriction of all philosophers. Hume explains the scope of the understanding in the final paragraphs of the section (EHU 12.26–34), which I discussed in the previous chapter.

Pyrrhonian skepticism can help motivate us to abide within the sphere of experience-based reason.

> To bring us to so salutary a determination, nothing can be more serviceable, than to be once thoroughly convinced of the force of the Pyrrhonian doubt, and of the impossibility, that any thing, but the strong power of natural instinct, could free us from it. . . . they [philosophers] will never be tempted to go beyond common life, so long as they consider the imperfection of those faculties which they employ, their narrow reach, and their inaccurate operations. While we cannot give a satisfactory reason, why we believe, after a thousand experiments, that a stone will fall, or fire burn; can we ever satisfy ourselves concerning any determination, which we may form, with regard to the origin of worlds, and the situation of nature, from, and to eternity? (EHU 12.25)

Pyrrhonian arguments provide the material for an a fortiori argument: if the beliefs of common life are unjustified, much more so are beliefs that go beyond common life. Hume does not say or imply that every argument put forward in support of extreme skepticism succeeds philosophically. Whether they philosophically succeed or not, they can motivate those who accept them to confine themselves to reasoning from experience.

Philo makes this same a fortiori argument in Part 1 of *Dialogues Concerning Natural Religion*:

> Let us become thoroughly sensible of the weakness, blindness, and narrow limits of human reason: Let us duly consider its uncertainty and endless contrari-

> eties, even in subjects of common life and practice: Let the errors and deceits of our very senses be set before us; the insuperable difficulties, which attend first principles in all systems; the contradictions, which adhere to the very ideas of matter, cause and effect, extension, space, time, motion; and in a word, quantity of all kinds, the object of the only science, that can fairly pretend to any certainty or evidence. When these topics are displayed in their full light, as they are by some philosophers and almost all divines; who can retain such confidence in this frail faculty of reason as to pay any regard to its determinations in points so sublime, so abstruse, so remote from common life and experience? When the coherence of the parts of a stone, or even that composition of parts, which renders it extended; when these familiar objects, I say, are so inexplicable, and contain circumstances so repugnant and contradictory; with what assurance can we decide concerning the origin of worlds, or trace their history from eternity to eternity? (DNR 1.3)

If human faculties do not produce justified beliefs about matters within the sphere of common life, much less do they produce justified conclusions about matters outside the sphere of common life. If our opinions grounded in common life are suspect, then opinions not grounded in common life are ludicrous.

Hume may not endorse all of the skeptical problems that Philo proposes here. These arguments may function as *reductio* skeptical arguments. For example, he mentions again "the contradictions, which adhere to the very ideas of . . . quantity of all kinds." Hume may take himself to have a coherent concept of quantity, as he suggests at in *Enquiry* 12, footnote 34. But, as in the *Enquiry*, it is hard to avoid the overall impression that Philo, and Hume, do think that some of these skeptical arguments are sound. In any case, if we accept any skeptical arguments within the sphere of common life, those will motivate total agnosticism about matters beyond the sphere of common life.

Qu thinks that the most significant advantage of his interpretation over skeptical ones is that "my interpretation offers a philosophical explanation for Hume's rationale to limit the scope of our enquiries. This is in contrast to sceptical interpretations, which are typically unable to explain Hume's rationale for this claim."[26] Qu's rationale is that the scope restriction limits us to the reliable operations of our faculties.[27] Hume's naturalistic study of the mind has yielded the empirical result that these operations are reliable and hence justified.

But the a fortiori argument explained above provides a perfectly respectable interpretation of how Pyrrhonism can motivate the scope restriction. If beliefs within this scope are epistemologically unjustified, then beliefs

26. Ibid., 209
27. Ibid., 210–11

outside this scope must be ludicrous. The a fortiori argument does presuppose a substantive claim, that beliefs outside the scope of common life must somehow have a lower epistemic standing than those within common life. But that seems plausible for Hume to take on board as a pretheoretical commitment.

The main point of Section 12 of the *Enquiry* is to motivate a moderate commitment to the demands of philosophy. Fanatical devotion to philosophy is a live option only if the demands of philosophy are always psychologically sustainable and practically beneficial. But skeptical arguments show that the demands of philosophy are not always sustainable or beneficial. With the Academics, we ought to demur from these demands when necessary. This does not open the door to unhinged speculation or superstitious revelation claims, however. Appreciation of the philosophical force of skepticism even within the bounds of reason and common life gives us an added motivation never to venture into the even more dubious territory beyond common life.

Chapter Eight

Skepticism, Irreligion, and Moderation

According to many interpreters, Hume does not want to end up as an epistemic skeptic. As I discussed in the Introduction, the weightiest argument for non-skeptical readings is that epistemic skepticism is incompatible with science, and Hume wants to do science, so he must not intend to end up as an epistemic skeptic. Hsueh Qu, for example, admits that the Title Principle is (in his view) a flawed epistemic solution to skepticism; however, it is more charitable to ascribe to Hume a bad epistemic solution than none at all.[1] According to Peter Millican, Hume stumbles unwillingly into epistemic skepticism in the *Treatise* and finds no way out.[2]

But I have argued that Hume not only puts forward skeptical arguments that he himself regards as sound, but also *reductio* arguments whose premises he does not himself endorse, in order to drive readers to a skeptical conclusion. He is not unwillingly forced into epistemic skepticism by intellectual integrity. He drives readers to skepticism by all available means.

Skeptical interpreters have not offered robust explanations of why Hume welcomes the philosophical (not practical) triumph of Pyrrhonism. Even if Hume's practical response to skepticism is theoretically viable, why does he consider it desirable, preferable to a non-skeptical outcome? Richard H. Popkin writes, "In being entirely the product of nature he welds his schizophrenic personality and philosophy together."[3] I have argued that Hume thinks his weld can hold together a moderate commitment to philosophy with the

1. Qu, *Hume's Epistemological Evolution*, 144n46
2. Millican, "Hume's Chief Argument," 95–98. According to Millican, Hume changes his ways and avoids epistemic skepticism in works after the *Treatise*. 98–101.
3. Richard H. Popkin, "David Hume: His Pyrrhonism and His Critique of Pyrrhonism." In *Hume: A Collection of Essays*, edited by V. C. Chappell, 95. Garden City, NY: Anchor Books, 1966. Reprinted from *The Philosophical Quarterly*, Vol. I (1951).

demands of common life. But why would he want to use epistemic skepticism to convince us that these two things sometimes diverge? Interpreters often point out that the skeptical crisis motivates intellectual modesty and remaining within the narrow bounds of the understanding (EHU 12.24–25). But Hume never says that a skeptical crisis is a necessary condition for adopting these attitudes. He advocates them before and apart from his skeptical arguments. They are features of normative philosophy *simpliciter*, not unique features of Academic philosophy.

In this chapter I suggest two reasons why Hume might welcome a radical skeptical conclusion to which the only response is purely practical. First, his radical skepticism coheres with the moderate approach to philosophy that he adopted in the wake of his youthful breakdown. Second, his radical skepticism deeply coheres with his irreligious aims, as a comparison with Descartes shows.

HUME'S YOUTHFUL BURNOUT AND SUBSEQUENT MODERATION

Biographical evidence indicates that Hume's youthful breakdown in September 1729 led him to adopt a moderate approach to philosophy. After a period of total commitment to his philosophical work, Hume suddenly found himself listless. The only way to recover his intellectual energy was to lead a more balanced life. It is plausible to think that Hume found in philosophical skepticism a further confirmation of a lesson he had learned by experience: that there are good practical reasons to sometimes demur from the demands of philosophy.

He describes his period of ecstatic philosophical devotion in his 1734 letter to a physician:

> After much Study, and Reflection on this [new Medium, by which Truth might be establish], at last, when I was about 18 Years of Age, there seem'd to be open'd up to me a new Scene of Thought, which transported me beyond Measure, and made me, with an Ardor natural to young men, throw up every other Pleasure or Business to apply entirely to it. The Law which was the Business I design'd to follow, appear'd nauseous to me, and I cou'd think of no other way of pushing my Fortune in the World, but that of a Scholar and Philosopher. I was infinitely happy in this Course of Life for some Months; till at last, about the beginning of Sept[l] 1729, all my Ardor seem'd in a moment to be extinguish, and I cou'd no longer raise my Mind to that pitch, which formerly gave me such excessive Pleasure. (HL i. 13)

Hume describes his experience in superlative terms: "transported . . . beyond Measure"; "infinitely happy"; "excessive Pleasure." The pleasures of Hume's studies surpassed moderation, by his own account. Moreover, he responded to his transporting experience by pursuing nothing but philosophy, exclusively and relentlessly—until his breakdown. He explicitly compares his own case to the case of religious enthusiasts.[4] At the time of the writing of the letter, he was still not fully out of the woods: "I found that I was not able to follow out any Train of Thought, by one continued Stretch of View, but by repeated Interruptions" (ibid, 16–17).

He learned through this experience that philosophy must be pursued in moderation, not in monomaniacal excess. He reports that between April 1730 and May 1731, among other measures taken for his health he took to riding regularly and moderated his studies, to good effect (HL i. 14–15). It is plausible to think that these experiences lie back of his recognition of the liabilities of philosophy in Section 1 of the first *Enquiry*, and his determination to pursue philosophy only when he feels like it in *Treatise* 1.4.7.11–15. Scholars often speculate about the emotional impact of Hume's youthful loss of religious faith, and I do not wish to discount that. But the available evidence indicates that the god that failed the teenage Hume may have been philosophy itself.

In light of this early crisis and his considered response to it, it is unsurprising to find Hume write in 1737 to his cousin, Henry Home, Lord Kames that "I was resolved not to be an enthusiast in philosophy, while I was blaming other enthusiasms" (HL i. 25). The statement is intended as an explanation of why "I am at present castrating my work [the *Treatise*], that is, cutting off its nobler parts; that is, endeavouring it shall give as little offence as possible, before which, I could not pretend to put it into the Doctor's [Dr. Joseph Butler's] hands" (ibid). Hume was hoping that Butler might endorse the *Treatise* and thereby bring it to the attention of the literate public.[5] He was presumably cutting out parts of the *Treatise* that impugned religious belief—perhaps his argument against miracles, or his refutation of the analogical design argument.

Why would sending the manuscript to Butler with the offensive parts included have made Hume "an enthusiast in philosophy"? To retain the offensive parts would have made it unlikely that Butler would endorse the

4. "I have notic'd in the Writings of the French Mysticks, and in those of our Fanatics here, that, when they give a History of the Situation of their Souls, they mention a Coldness and Desertion of the Spirit, which frequently returns, and some of them, at the beginning, have been tormented with it many Years. As this kind of Devotion depends entirely on the Force of Passion, and consequently of the Animal Spirits, I have often thought that their Case and mine were pretty parallel, and that their rapturous Admirations might discompose the Fabric of the Nerves and Brain, as much as profound Reflections, and that warmth or Enthusiasm which is inseperable from them" (HL i. 17).

5. E. C. Mossner, *The Life of David Hume*, 2nd ed. (New York: Clarendon Press, 1980), 110–12; Harris, *Hume*, 116–17.

book, and therefore would have made it less likely that the *Treatise* would get a wide readership. In short, it would have defeated the whole purpose of publishing his philosophy in the first place. Hume's "castration" of the book perhaps reflects his realization that philosophy should never be pursued to one's own harm.

The ideal of moderation seems to have guided Hume's life, at least after the 1729 crisis. In *The Natural History of Religion*, he writes that "In general, no course of life has such safety (for happiness is not to be dreamed of) as the temperate and moderate, which maintains, as far as possible, a mediocrity, and a kind of insensibility, in every thing" (NHR 15.3). In "My Own Life," Hume describes himself essentially in terms of moderation: "I was, I say, a man of mild disposition, of command of temper, of an open, social, and cheerful humour, capable of attachment, but little susceptible of enmity, and of great moderation in all my passions" (MOL xl).

Adopting a moderate attitude toward the demands of philosophy coheres well with the approach to life Hume describes in these passages. If Hume wished to commend a moderate life to fanatical devotees of philosophy, then it would serve his purposes quite well to persuade them that such fanaticism has fatal skeptical consequences. He would have a good reason to put forward the skeptical arguments that he himself regards as sound, and *reductio* skeptical arguments as well. Either kind of argument would lead someone to the true recognition that they should moderate their commitment to the demands of philosophy.

SKEPTICISM AND A NON-DECEIVING GOD: HUME VERSUS DESCARTES

Hume frames his response to skepticism as the opposite of Descartes's. The despairing first-person narrative of *Treatise* 1.4.7 clearly invites comparison with the First Meditation. Descartes's skeptical doubts about the external world and the reliability of reason come in the First Meditation, at the outset of his investigations. The First Meditation closes amid "the inextricable darkness of the problems I have now raised."[6] The Second Meditation opens,

> So serious are the doubts into which I have been thrown as a result of yesterday's meditation that I can neither put them out of my mind nor see any way of resolving them. It feels as if I have fallen unexpectedly into a deep whirlpool

6. René Descartes, *Meditations on First Philosophy*, in *The Philosophical Writings of Descartes, Volume II*, translated by John Cottingham, Robert Stoothoff, and Dugald Murdoch (New York: Cambridge University Press, 1984), 15.

> which tumbles me around so that I can neither stand on the bottom nor swim up to the top.[7]

Hume writes,

> The *intense* view of these manifold contradictions and imperfections in human reason has so wrought upon me, and heated my brain, that I am ready to reject all belief and reasoning, and can look upon no opinion even as more probable or likely than another. Where am I, or what? From what causes do I derive my existence, and to what condition shall I return? Whose favour shall I court, and whose anger must I dread? What beings surround me? and on whom have I any influence, or who have any influence on me? I am confounded with all these questions, and begin to fancy myself in the most deplorable condition imaginable, inviron'd with the deepest darkness, and utterly depriv'd of the use of every member and faculty. (T 1.4.7.8)

Both Descartes and Hume recognize that although their ordinary beliefs are unjustified, they are psychologically difficult if not impossible to resist. Descartes responds by doing everything he can to resist these unjustified beliefs. As a remedy, he resolves to overcompensate by treating his habitual (unjustified) beliefs as not merely uncertain but false. "I shall stubbornly and firmly persist in this meditation; and, even if it is not in my power to know any truth, I shall at least do what is in my power, that is, resolutely guard against assenting to any falsehoods."[8] He sees no practical dangers in this plan, "because the task now in hand does not involve action but merely the acquisition of knowledge." Descartes remains, as it were, in his study, "quite alone."[9] Hume, by contrast, welcomes the irresistible psychological force of his natural beliefs.

> Most fortunately it happens, that since reason is incapable of dispelling these clouds, nature herself suffices to that purpose, and cures me of this philosophical melancholy and delirium, either by relaxing this bent of mind, or be some avocation, and lively impression of my senses, which obliterate all these chimeras. I dine, I play a game of back-gammon, I converse, and am merry with my friends; and when after three or four hour's amusement, I wou'd return to these speculations, they appear so cold, and strain'd and ridiculous, that I cannot find in my heart to enter into them any farther. (T 1.4.7.9)

He finds absolutely no practical reason to "strive against the current of nature, which leads me to indolence and pleasure" (T 1.4.7.10). Descartes, by contrast, strenuously resists "a kind of laziness" that "brings me back to normal life."[10]

7. Ibid., 16.
8. Ibid., 15.
9. Ibid., 12.
10. Ibid., 15.

Descartes adheres ruthlessly to the demands of philosophy in the hope, if not the faith, that doing so will lead to certainty and truth. That hope comes to fruition when he proves the existence of a non-deceiving God who guarantees the reliability of his faculties when used properly. As it turns out, error is not the fault of our God-given faculties, but results from our culpable misuse of those faculties. Adhering to the demands of philosophy, and properly using our reason, turns out to be an act of obedience to God which he rewards with the attainment of truth and certainty.

Hume has no reason to hope or expect that adherence to the demands of philosophy will be rewarded with certainty and truth. Unlike Descartes, he has already exhausted his intellectual options. Descartes begins in skeptical doubt, but ascends through argument to certainty. Hume begins the *Treatise* with confidence but descends through argument to skeptical doubt. He explicitly contrasts Descartes's "antecedent skepticism" with his own "consequent skepticism" in the *Enquiry* (EHU 12.3, 5). Antecedent skepticism may be more demanding in some ways, but consequent skepticism is more final.

Just as Descartes uses theism to conquer skepticism, Hume uses skepticism to conquer theism. First, Hume's skepticism about the external world cuts off all possible theistic proofs. In a paragraph clearly aimed at Descartes, Hume says that knowledge of the world must precede any claims about a Supreme Being: "if the external world once be called in question, we shall be at a loss to prove the existence of that Being or any of his attributes" (EHU 12.13). In the *Dialogues*, Cleanthes points out that a total philosophical skepticism is a threat to natural theologians like Locke who would base all religious claims on philosophical argument (DNR 1.17).[11] However, except in these passages, Hume does not invoke this skeptical argument against natural theology. As an Academic skeptic, he accepts beliefs grounded in common life on purely practical grounds. This leaves open the possibility, in theory, of religious arguments grounded in common life.

Secondly, Hume uses philosophical skepticism as an argument against theism. The defects that philosophical skepticism exposes in our faculties, even at their best and when used properly, provides material for an argument against the existence of a non-deceiving creator. Descartes presents just this argument in the First Meditation: "But if it were inconsistent with his goodness to have created me such that I am deceived all the time, it would seem equally foreign to his goodness to allow me to be deceived even occasionally; yet this last assertion cannot be made."[12] Hume repeats the same argument when he says "If his [the Supreme Being's] veracity were at all concerned in this matter [the veracity of our senses], our senses would be entirely infallible; because it is not possible that he can ever deceive" (EHU 12.13). Descartes

11. Paul Russell, *Riddle*, 213
12. Descartes, *Meditations*, 14.

has an answer: our faculties lead to error only when we misuse them (Fourth Meditation). However, Hume's skeptical arguments against the senses show that even when we employ our faculties correctly, according to the norms of philosophy, they do not lead to true beliefs but to total suspension of belief in the external world.[13] Descartes holds that since there is a non-deceiving God, total commitment to the demands of philosophy (to the proper use of our faculties) is warranted; philosophy will lead us to truth, at least in the long run. Hume can argue that that since total commitment to philosophy does not lead to truth but to skepticism, there is no non-deceiving God.

In keeping with this negative conclusion about the causes of the human mind, Hume stresses its defects in the key chapters on skepticism. In the *Treatise*, he compares his cognitive faculties to a "leaky weather-beaten vessel" (T 1.4.7.1). He bemoans "The wretched condition, weakness, and disorder of the faculties, I must employ in my enquiries" (ibid.). He finds in them "so many [infirmities] which are common to human nature" (T 1.4.7.2). A non-skeptical interpreter might object that these are Hume's sentiments before he develops his response to skepticism, and that he recovers his confidence later. But in the *Enquiry*, Hume makes perhaps even harsher statements about our faculties even after rejecting Pyrrhonism. He points out "the strange infirmities of the human understanding, even in its most perfect state, and when most accurate and cautious in its determinations" (EHU 12.24). "Universal perplexity and confusion . . . is inherent in human nature" (ibid). Academic philosophers are those who take to heart "the imperfection of those faculties which they employ, their narrow reach, and their inaccurate operations" (EHU 12.25). Philo speaks of "the weakness, blindness, and narrow limits of human reason . . . its uncertainty and endless contrarieties, even in subjects of common life and practice . . . this frail faculty of reason" (DNR 1.3).

Against Descartes, Hume replies that since there is no God to guarantee that philosophy will eventually lead us to practically livable beliefs, we had better just leave our study and have dinner. If we stay locked up, desperately trying to suspend our natural beliefs until we find some justification for them, we will starve to death. For Hume, we are here merely by chance, and Pyrrhonism has the virtue of highlighting "the whimsical condition of mankind" (EHU 12.23). Hume's philosophical skepticism supports his irreligious aims; his practical response shows us how to live with cognitive faculties that do not come with a divine warranty.

Someone might object that this argument against religion is unnecessary and self-defeating. Hume has plenty of powerful irreligious arguments that presuppose the epistemic authority of our faculties. He does not need to make one that presupposes skepticism. Moreover, skepticism undercuts his other

---

13. Hume would also object to the libertarian account of free will on which Descartes's theodicy in the Fourth Meditation depends (cf. EHU 8, especially 8.36).

irreligious arguments. If none of our beliefs are justified, then none of our irreligious beliefs are justified either. But even if philosophical skepticism is unnecessary to disprove a non-deceiving creator, it is sufficient. Furthermore, Hume's other non-skeptical, reason-based arguments against religion serve a complementary role. Even if we agree to ignore general skeptical arguments, religion still cannot philosophically establish its claims.

Paul Russell also argues that Hume's skepticism has a fundamentally irreligious aim.[14] Russell writes, "The basic intent of Hume's skeptical principles and arguments, as here explained, is to discredit the metaphysical and epistemological ambitions of the 'religious philosophers,' while *at the same time* leaving some adequate foundation upon which to pursue his (Hobbist) project of a 'science of man.'"[15] Russell however does not identify the particular anti-religious argument that I find by comparing Hume with Descartes. Instead, Russell claims that by limiting philosophical inquiry to common life, Hume cuts off theological speculations.[16]

The scope limitation, however, is neither necessary nor sufficient to achieve Hume's irreligious aims. The scope limitation is a feature of philosophy *simpliciter*. It is not unique to skepticism, whether Pyrrhonian or Academic. The skeptic simply has an added motive to abide by this scope limitation. Furthermore, the scope limitation does not, by itself, rule out religious conclusions. It does rule out the kind of religious skepticism, or fideism, that relies on mere faith, unsupported by reason or philosophy.[17] But the scope limitation does not prevent the natural theologian from offering rational arguments from experience to the existence of God. Hume acknowledges that such arguments are, in principle, legitimate:

> Divinity or Theology, as it proves the existence of a Deity, and the immortality of souls, is composed partly of reasonings concerning particular, partly concerning general facts. It has a foundation in *reason*, so far as it is supported by experience. (EHU 12.32)

The problem is just that experimental arguments for the deity and for miracles all fail, as Hume shows in Sections 10–11 (and in the *Dialogues*).[18] When he says, "But its [Theology's] best and most solid foundation is *faith* and

14. Russell, *Riddle*, chapter 15, "The Practical Pyrrhonist," 204–222.

15. Ibid. 222

16. Ibid. 220

17. Hume sometimes gives the impression that he might be this kind skeptic, as in the *Letter to a Gentleman*. See Russell's discussion. Philo makes the same gambit in Part 1 of the *Dialogues*.

18. As Russell notes, 209. Cleanthes argues that it would be inconsistent for skeptics like Philo to accept, as they do, even abstruse arguments for scientific conclusions, but dismiss without examination the arguments of natural theology (DNR 1.12–16). Philo makes no objection, and the ensuing discussion in the rest of the book indicates that he agrees with Cleanthes. He rejects the arguments of natural theology, not because of their subject matter, but because they are bad arguments.

divine revelation" (EHU 12.32), he ironically implies that it has no basis in experience-based reasoning. So neither is skepticism necessary to motivate the scope limitation, nor is the scope limitation sufficient to rule out religion.

Hume wants to arrive at the conclusion of epistemic skepticism because it motivates us to adopt a moderate attitude toward philosophy. If philosophy led only onward and upward to a Grand Unified Theory of Everything, then an absolute commitment to this method of inquiry would seem practically justified. The Pyrrhonian skeptical crisis highlights the sad end to which a violent passion for philosophy must come.

Bibliography

Ainslie, Donald C. *Hume's True Scepticism*. New York: Oxford University Press, 2015.

Allison, Henry. "Hume's Philosophical Insouciance: A Reading of *Treatise* 1.4.7," *Hume Studies* 31 (2005), 317–46.

Alston, William P. *Beyond "Justification": Dimensions of Epistemic Evaluation*. Ithaca, NY: Cornell University Press, 2005.

———. *The Reliability of Sense Perception*. Ithaca: Cornell University Press, 1993.

Audi, Robert. *Epistemology: A Contemporary Introduction to the Theory of Knowledge*, 2nd ed. New York: Routledge, 2003.

Baier, Annette C. *A Progress of Sentiments: Reflections on Hume's* Treatise. Cambridge, MA: Harvard University Press, 1991.

Bailey, Alan, and Dan O'Brien, eds. *The Continuum Companion to Hume*. New York: Continuum, 2012.

Baxter, Donald L. M. "Hume, Distinctions of Reason, and Differential Resemblance," *Philosophy and Phenomenological Research* 82.1 (January 2011), 156–82.

———. *Hume's Difficulty: Time and Identity in the* Treatise (New York: Routledge, 2008).

Broughton, Janet. "Hume's Naturalism and His Skepticism." In *A Companion to Hume*, ed. Elizabeth S. Radcliffe, 425–40. Malden, MA: Blackwell, 2008.

———. "Hume's Skepticism about Causal Inferences." *Pacific Philosophical Quarterly* 64, 1983, 3–18.

———. "The Inquiry in the *Treatise*." *Philosophical Review*, 113, no. 4 (November 2004): 537–56.

Buckle, Stephen. *Hume's Enlightenment Tract: The Unity and Purpose of* An Enquiry concerning Human Understanding (Oxford: Clarendon Press, 2001).

Butler, Annemarie. "Hume on Believing the Vulgar Fiction of Continued Existence." *History of Philosophy Quarterly* 27, no. 3 (2010): 237–54.

———. "Vulgar Habits and Hume's Double Vision Argument." *The Journal of Scottish Philosophy* 8, no. 2 (2010): 169–87.

Chignell, Andrew. "The Ethics of Belief." *The Stanford Encyclopedia of Philosophy*, Spring 2013 edition, ed. Edward N. Zalta, accessed November 30, 2015, http://plato.stanford.edu/archives/spr2013/entries/ethics-belief.

Clifford, W. K. "The Ethics of Belief," *Contemporary Review*, January 1877; reprinted in *Lectures and Essays*, ed. Leslie Stephen and Frederick Pollock (London: Macmillan and Co., 1886), 339–63.

Cummins, Philip. "Hume's Diffident Skepticism," *Hume Studies* 25, no. 1/2 (April/November 1999): 43–65.

De Pierris, Graciela. "Hume's Pyrrhonian Skepticism and the Belief in Causal Laws." *Journal of the History of Philosophy* 39, no. 3 (July 2001): 351–83.

———. *Ideas, Evidence, and Method: Hume's Skepticism and Naturalism Concerning Knowledge and Causation*. New York: Oxford University Press, 2015.

Descartes, René. *The Philosophical Writings of Descartes, Volume II*. Translated by John Cottingham, Robert Stoothoff, and Dugald Murdoch. New York: Cambridge University Press, 1984.

Durland, Karann. "Extreme Skepticism and Commitment in the *Treatise*." *Hume Studies* 37, no. 1 (April 2011): 65–98.

Fogelin, Robert J. *Hume's Skeptical Crisis*. New York: Oxford University Press, 2009.

———. "Hume's Skepticism." In *The Cambridge Companion to Hume*, 2nd ed., edited by David Fate Norton and Jacqueline Taylor, 209–237. Cambridge: Cambridge University Press, 2009.

———. *Hume's Skepticism in the Treatise of Human Nature*. Boston: Routledge and Kegan Paul, 1985.

Frasca-Spada, Marina. *Space and Self in Hume's* Treatise. New York: Cambridge University Press, 1998.

Garrett, Don. *Cognition and Commitment in Hume's Philosophy*. New York: Oxford University Press, 1997.

———. "Introduction." In Norman Kemp Smith, *The Philosophy of David Hume*. New York: Palgrave Macmillan, 1941, 2005.

———. *Hume*. New York: Routledge, 2015.

———. "Hume's Conclusions in 'Conclusion of this Book.'" In *The Blackwell Guide to Hume's* Treatise, edited by Saul Traiger, 151–75. Malden, MA: Blackwell, 2006.

———. "Hume's Theory of Causation: Inference, Judgment, and the Causal Sense," in *The Cambridge Companion to Hume's* Treatise, edited by Donald C. Ainslie and Annemarie Butler (New York: Cambridge University Press, 2015), 69–100.

———. "Millican's 'Abstract,' 'Imaginative,' 'Reasonable,' and 'Sensible' Questions about Hume's Theory of Cognition," in *Hume Studies* vol. 40, no. 2 (November 2014): 227–42.

Greco, John. "Virtues in Epistemology." In *The Oxford Handbook of Epistemology*, edited by Paul K. Moser, 287–315. Oxford: Oxford University Press, 2002.

Green, T. H. *Hume and Locke*. New York: Thomas Y. Crowell Co., 1968.

Hacking, Ian. "Hume's Species of Probability." *Philosophical Studies* 33, no. 1 (January 1978): 21–37.

Harris, James. *Hume: An Intellectual Biography* (New York: Cambridge University Press, 2015).

———. "Introduction." In *The Oxford Handbook of British Philosophy in the Eighteenth Century*, edited by James A. Harris, 1–17. New York: Oxford University Press, 2013.

Hume, David. *An Abstract of a Book Lately Published; Entituled, A Treatise of Human Nature, &c. Wherein the Chief Argument of That Book is Farther Illustrated and Explained*, at *Hume Texts Online* (https://davidhume.org), eds. Amyas Merivale and Peter Millican. Citations indicate paragraph numbers.

———. *An Enquiry Concerning Human Understanding*, at *Hume Texts Online* (https://davidhume.org), eds. Amyas Merivale and Peter Millican.

———. *An Enquiry Concerning the Principles of Morals*, edited by Tom L. Beauchamp. Oxford: Oxford University Press, 1998.

———. *Dialogues Concerning Natural Religion and Other Writings*, ed. Dorothy Coleman, Cambridge: Cambridge University Press, 2007.

———. *Dissertation on the Passions; The Natural History of Religion: A Critical Edition*, edited by Tom L. Beauchamp. Oxford: Clarendon Press, 2007.

———. *Essays Moral, Political, Literary*, ed. Eugene F. Miller. Indianapolis: Liberty Fund, 1987.

———. *Letter from a Gentleman to His Friend in Edinburgh*, edited by Ernest C. Mossner and John V. Price, Edinburgh: Edinburgh University Press, 1967.

———. *Letters of David Hume*, 2 vols., edited by J. Y. T. Greig. New York: Oxford University Press, 1932.

———. "My Own Life." In *Essays Moral, Political, Literary*, edited by Eugene F. Miller. Indianapolis: Liberty Fund, 1987.

———. *Treatise of Human Nature*, at *Hume Texts Online* (https://davidhume.org), eds. Amyas Merivale and Peter Millican.

Imlay, Robert. "Hume's 'Of scepticism with regard to reason,': A Study in Contrasting Themes." *Hume Studies* vol. 7, no. 2 (November 1981): 121–36.

Johnson, Oliver. *The Mind of David Hume: A Companion to Book I of* A Treatise of Human Nature. Urbana and Chicago: University of Illinois Press, 1995.

Kail, P. J. E. *Projection and Realism in Hume's Philosophy*. New York: Oxford University Press, 2007.

Kant, Immanuel. *Prolegomena to Any Future Metaphysics*, translated by Gary Hatfield, in *Theoretical Philosophy after 1781*, edited by Henry Allison and Peter Heath. New York: Cambridge University Press, 2002.

Karlsson, Mikael M. "Epistemic Leaks and Epistemic Meltdowns: A Response to William Morris on Scepticism with Regard to Reason," *Hume Studies* vol. 16, no. 2 (1990), 121–30.

Kelly, Thomas. "Evidence," in *The Stanford Encyclopedia of Philosophy* Fall 2014 ed., edited by Edward N. Zalta, accessed November 30, 2015, http://plato.stanford.edu/archives/fall2014/entries/evidence/

Kemp Smith, Norman. 'The Naturalism of Hume (I).' *Mind* 14, no. 54 (1905), 149–73.

———. *The Philosophy of David Hume*, New York: Palgrave Macmillan, 2005.

Locke, John. *An Essay concerning Human Understanding*, edited by Peter H. Nidditch. Oxford: Clarendon Press, 1975.

Loeb, Louis E. *Stability and Justification in Hume's Treatise*. New York: Oxford University Press, 2002.

Lynch, Michael P. "Hume and the Limits of Reason." *Hume Studies* vol. 22, no. 1 (April 1996), 89–104.

Lyons, Jack C. "General Rules and the Justification of Probable Belief in Hume's *Treatise*." *Hume Studies* 27, no. 2(November 2001), 247–78.

Martin, Marie A. "The Rational Warrant for Hume's General Rules." *Journal of the History of Philosophy* 31, no. 2 (1993), 245–57.

Meeker, Kevin. "Hume's Iterative Probability Argument: A Pernicious *Reductio*." *Journal of the History of Philosophy* 38, no. 2 (April 2000), 221–38.

———. "Hume: Radical Skeptic or Naturalized Epistemologist?" *Hume Studies* vol. 24, no. 1 (April 1998), 31–52.

———. *Hume's Radical Scepticism and the Fate of Naturalized Epistemology*. New York: Palgrave Macmillan, 2013.

Millican, Peter. "Hume's Sceptical Doubts concerning Induction." In *Reading Hume on Human Understanding*, edited by Peter Millican, 107–174. New York: Oxford University Press, 2002.

———. "Hume's Chief Argument." In *The Oxford Handbook of Hume*, edited by Paul Russell, 82–108. New York: Oxford University Press, 2016.

———. "Hume's 'Scepticism' About Induction." In *The Continuum Companion to Hume*, edited by Alan Bailey and Dan O'Brien, 57–103. New York: Continuum, 2012.

———. "Skepticism about Garrett's *Hume*: Faculties, Concepts, and Imposed Coherence," in *Hume Studies* vol. 40, no. 2 (November 2014), 205–226.

Morris, William Edward. "Belief, Probability, Normativity." In *The Blackwell Guide to Hume's Treatise*, edited by Saul Traiger, 77–94. Malden, MA: Blackwell Publishing, 2006.

———. "Hume's Conclusion," *Philosophical Studies* 99 (2000), 89–110.

———. "Hume's Scepticism About Reason" *Hume Studies* 15, no. 1 (April 1989), 40–53.

Mossner, E. C. *The Life of David Hume*, 2nd ed. New York: Clarendon Press, 1980.

Mounce, H. O. *Hume's Naturalism*. New York: Routledge, 1999.

Noonan, Harold W. *Hume on Knowledge*. New York: Routledge, 1999.

Norton, David Fate. "Of the Academical or Sceptical Philosophy." In *Reading Hume on Human Understanding*, edited by Peter Millican, 371–92. Oxford: Clarendon Press, 2002.

Owen, David. *Hume's Reason*. New York: Oxford University Press, 1999.

———. "Scepticism with Regard to Reason." In *The Cambridge Companion to Hume*, edited by Donald C. Ainsle and Annemarie Butler, 101–134. New York: Cambridge University Press, 2015.

Plantinga, Alvin. *Warrant and Proper Function*. Oxford: Oxford University Press, 1993.

Pollock, John. *Contemporary Theories of Knowledge*. Towota, NJ: Rowman & Littlefield, 1986.

Popkin, Richard H. "David Hume: His Pyrrhonism and His Critique of Pyrrhonism." In *Hume: A Collection of Essays*, edited by V. C. Chappell, 53–98. Garden City, NY: Anchor Books, 1966. Reprinted from *The Philosophical Quarterly*, Vol. 1 (1951).

Qu, Hsueh M. *Hume's Epistemological Evolution*. New York: Oxford University Press, 2020.

———. "Hume's Practically Epistemic Conclusions?" *Philosophical Studies* 170, no. 3 (September 2014), 501–524.

Quine, W.V. "Epistemology Naturalized." In *Ontological Relativity and Other Essays*, 69–90. New York: Columbia University Press, 1969.

Reid, Thomas. *Essays on the Intellectual Powers of Man*. Cambridge, MA: MIT Press, 1969.

———. *Inquiry into the Human Mind*, in *The Works of Thomas Reid*, vol. 1, 7th ed., edited by William Hamilton. Edinburgh: Elibron Classics, 2005.

Ridge, Michael. "Epistemology Moralized: David Hume's Practical Epistemology." *Hume Studies* vol. 29, no. 2 (November 2003), 165–204.

Rocknak, Stefanie. *Imagined Causes: Hume's Conception of Objects*. New York: Springer, 2013.

Russell, Paul, ed. *The Oxford Handbook of Hume*. New York: Oxford University Press, 2016.

———. *The Riddle of Hume's* Treatise*: Skepticism, Naturalism, and Irreligion*. New York: Oxford University Press, 2008.

Rutherford, Donald. "Innovation and orthodoxy in early modern philosophy." In *The Cambridge Companion to Early Modern Philosophy*, edited by Donald Rutherford, 11–38. New York: Cambridge University Press, 2006.

Sasser, Nathan I. "Hume and the Implanted Knowledge of God," *Journal of Scottish Philosophy*, 13(1), March 2015, 17–35.

———. "Hume on the Defeasible Justification of the Vulgar Belief in Body," in *History of Philosophy Quarterly* 36.4 (October 2019), 359–76.

———. "Hume's Purely Practical Response to Philosophical Skepticism," in *Hume Studies* vol. 43, no. 2 (November 2017), 3–28.

Schafer, Karl. "Curious Virtues in Hume's Epistemology," *Philosophers' Imprint* 14, no. 1 (January 2014), 1–20.

Schmitt, Frederick F. *Hume's Epistemology in the* Treatise*: A Veritistic Interpretation*. New York: Oxford University Press, 2014.

Serjeantson, R. W. "Becoming a Philosopher in Seventeenth-Century Britain." In *The Oxford Handbook of Seventeenth-Century British Philosophy*, edited by Peter R. Anstey, 9–31. New York: Oxford University Press, 2013.

Singer, Ira. "Hume's extreme skepticism in *Treatise* I IV 7." *Canadian Journal of Philosophy* 25, no. 4 (December 2004), 595–622.

———. "Nature Breaks Down: Hume's Problematic Naturalism in *Treatise* I iv." *Hume Studies* vol. 26, no. 2 (November 2000), 225–43.

Stove, D. C. "Hume, Kemp Smith, and Carnap." *Australasian Journal of Philosophy* 55, no. 3 (December 1977), 189–200.

Stroud, Barry. *Hume.* New York: Routledge and Kegan Paul, 1977.

———. "Hume's Scepticism: Natural Instincts and Philosophical Reflection." In *The Empiricists: Critical Essays on Locke, Berkeley, and Hume*, edited by Margaret Atherton, 229–52. Lanham, MD: Rowman & Littlefield, 1999.

Williams, Michael. "Hume's Skepticism." In *The Oxford Handbook of Skepticism*, edited by John Greco, 80–107. Oxford: Oxford University Press, 2008.

Wilson, Fred. *The External World and our Knowledge of It: Hume's Critical Realism, an Exposition and a Defence*. Toronto: University of Toronto Press, 2008.

Winkler, Kenneth P. "Hume's Inductive Skepticism." In *The Empiricists: Critical Essays on Locke, Berkeley, and Hume*, edited by Margaret Atherton, 183–212. Lanham, MD: Rowman & Littlefield, 1999.

———. "Hume on Scepticism and the Senses." In *The Cambridge Companion to Hume's Treatise*, edited by Donald C. Ainslie and Annemarie Butler, 135–64. New York: Cambridge University Press, 2015.

———. "Hume's Skeptical Logic of Induction." In *The Oxford Handbook of Hume*, edited by Paul Russell, 91–227. New York: Oxford University Press, 2016.

Index

About the Author

Nathan I. Sasser is assistant professor of philosophy and Head of the Humanities Department at Greenville Technical College in Greenville, South Carolina. He teaches historically oriented courses in ethics and the problems of philosophy as well as logic and leadership. His research interests include David Hume, early modern philosophy, epistemology, and philosophy of religion. He has served as president of the South Carolina Society for Philosophy.

www.ingramcontent.com/pod-product-compliance
Lightning Source LLC
LaVergne TN
LVHW091530200926
848483LV00002B/354
9781793623218